# GRAPHING CALCULATOR MANUAL

## SANDRA DeGUZMAN
*Dutchess Community College*

# INTRODUCTION TO TECHNICAL MATHEMATICS
## FIFTH EDITION

## Allyn J. Washington
*Dutchess Community College*

## Mario F. Triola
*Dutchess Community College*

## Ellena E. Reda
*Dutchess Community College*

PEARSON
Addison
Wesley

Boston  San Francisco  New York
London  Toronto  Sydney  Tokyo  Singapore  Madrid
Mexico City  Munich  Paris  Cape Town  Hong Kong  Montreal

Copyright © 2008 Pearson Education, Inc.
Publishing as Pearson Addison-Wesley, 75 Arlington Street, Boston, MA 02116.

ISBN-13: 978-0-321-45062-3
ISBN-10: 0-321-45062-0

1 2 3 4 5 6 OPM 10 09 08 07

# Table of Contents

# Introduction

The TI-83, TI-83 Plus and the TI-84 Plus calculator will assist you in finding solutions to mathematical problems, allow you to check your hand computations and help you develop mastery of the material in your text, *Introduction to Basic Technical Mathematics/5e* by Washington/Triola/Reda. Most of the key strokes are identical for the three calculators. When the key strokes are different, the TI-83 key strokes are given first.

Included in each chapter are the step by step calculator instructions needed to solve some example problems taken from the main text. You can identify these problems in the manual by the bolded section and example number. These examples are aligned with the material you will be learning in each chapter of the text. Following many of the examples, are troubleshooting tips for common errors.

You will discover that you can solve various tasks on the calculator using a variety of methods. The manual shows at least one method, but you may find an equally effective method that works for you. You are encouraged to explore the calculator menus for future use.

As you approach your study of basic technical mathematics, it is recommended that you read each appropriate chapter and learn the necessary calculator key strokes. You will use many of the features on the graphing calculator for this course. Remember that the calculator is a wonderful tool but it is not a substitute for good critical thinking and problem solving skills.

# Chapter 1

# Signed Numbers

## Getting Started

Begin by familiarizing yourself with the calculator key pad for the TI-83/83 Plus/84 Plus. To turn the calculator on, press the [ON] key in the bottom left corner. You should see a flashing cursor on your screen.

If the cursor is faint, you can adjust the contrast on the screen by pressing [2nd] and then use the up arrow key to make it darker. (The [2nd] key is located in the upper left corner and the arrow keys are in the upper right.) You will see numbers flash in the upper right hand corner of the screen. They go up to 9. The higher the number the darker the screen should be. If you notice that your setting is approaching 9 and you are still having trouble viewing the screen, try changing the batteries.

To turn the calculator off, press [2nd] [OFF]. Whenever you see the action you want to take printed in yellow for the TI-83/83 Plus or blue for the TI-84 Plus, you must press the $2^{nd}$ key first. Since OFF is written in yellow/blue directly above the [ON] key, we pressed [2nd] [OFF] to turn the calculator off.

Any time you are in the main screen, you may press the [CLEAR] button located in the upper right side of the key pad if you want erase your screen. Later when you are in some of the more advanced screens, you may need to press [2nd] [QUIT] to quit and return to the main screen. Again notice the word QUIT is written in yellow/blue above the [MODE] button.

If you make a mistake and have already pressed [ENTER], you can recall the previous entry and edit it by pressing [2nd] [ENTRY]. Use the arrow keys to move the cursor to the mistake for correction. If you want to delete a number use the [DEL] key not [CLEAR]. If you want to insert another number or sign you use [2nd] [INS].

## Signed Numbers

In **Section 1.2, Example 8**, we must compute $(-8) + (+12)$. To do this computation on the calculator you must use the [(-)] key located on the bottom right hand corner. This key is used to indicate a negative number and is not a subtraction sign. You may enter positive signed numbers as unsigned numbers. We use the following key strokes [(-)] [8] [+] [1] [2] [ENTER] to get our result pictured on the next page.

Suppose that we want to solve 7-(-14). Here we use the subtraction key located on the right side of the calculator for the first minus sign and then the $\boxed{(-)}$ key for the sign directly in front of the 14. The key strokes are $\boxed{7}$ $\boxed{-}$ $\boxed{(-)}$ $\boxed{1}$ $\boxed{4}$ $\boxed{\text{ENTER}}$.

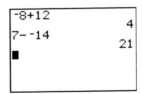

Note that the negative key $\boxed{(-)}$ appears on your screen as a smaller sign and slightly higher than the subtraction key.

Further in **Section 1.3, Example 3** we are asked to multiply $(-2)\times(+3)\times(+5)$, $(-2)\times(-3)\times(+5)$ and $(-2)\times(-3)\times(-5)$. The first one is entered as $\boxed{(-)}$ $\boxed{2}$ $\boxed{\times}$ $\boxed{3}$ $\boxed{\times}$ $\boxed{5}$ $\boxed{\text{ENTER}}$. The second one is entered $\boxed{(-)}$ $\boxed{2}$ $\boxed{\times}$ $\boxed{(-)}$ $\boxed{3}$ $\boxed{\times}$ $\boxed{5}$ $\boxed{\text{ENTER}}$. Finally, the third one is entered as $\boxed{(-)}$ $\boxed{2}$ $\boxed{\times}$ $\boxed{(-)}$ $\boxed{3}$ $\boxed{\times}$ $\boxed{(-)}$ $\boxed{5}$ $\boxed{\text{ENTER}}$.

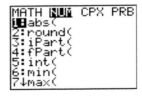

Again note the smaller negative sign is used for signed numbers. If you use $\boxed{-}$ you will get an error.

## Absolute Value and Using Menus

Absolute value is generally indicated by | | around the number being evaluated. On the calculator it is indicated by abs() where the number goes in the parentheses. To access the absolute value function on the graphing calculator, we must use the $\boxed{\text{MATH}}$ menu.

When you press the $\boxed{\text{MATH}}$ key located 3 keys down from the top on the left, you will see four categories at the top of your screen; MATH, NUM, CPX, PRB. Absolute value is found in the NUM category. Next use the $\boxed{\blacktriangleright}$ key on the key pad in the upper right to cursor over to NUM. It should be highlighted as shown in the picture below. Choose $\boxed{1}$ for the abs() menu item. Instead of pressing the number of your choice, you may use the up and down arrow key to highlight the menu item you want and then push $\boxed{\text{ENTER}}$.

In **Section 1.1, Example 6**, we are asked to find $\left|-\dfrac{3}{2}\right|$, $|-7.3|$, and $|9.2|$.

The key strokes for the first absolute value, $\left|-\dfrac{3}{2}\right|$, are $\boxed{\text{MATH}}$ $\boxed{\blacktriangleright}$ $\boxed{1}$ $\boxed{(\text{-})}$ $\boxed{3}$ $\boxed{\div}$ $\boxed{2}$ $\boxed{)}$ $\boxed{\text{MATH}}$ $\boxed{1}$
$\boxed{\text{ENTER}}$. Note that to get the answer displayed as a fraction we used the additional symbol
$\blacktriangleright$ Frac accessed with the keys $\boxed{\text{MATH}}$ $\boxed{1}$ at the end of our key strokes. If you exclude that
command the calculator will return 1.5 which is equivalent.

The second absolute value, $|-7.3|$, is entered using $\boxed{\text{MATH}}$ $\boxed{\blacktriangleright}$ $\boxed{1}$ $\boxed{(\text{-})}$ $\boxed{7}$ $\boxed{.}$ $\boxed{3}$ $\boxed{)}$ $\boxed{\text{ENTER}}$.

Finally, the third absolute value, $|9.2|$, is found by $\boxed{\text{MATH}}$ $\boxed{\blacktriangleright}$ $\boxed{1}$ $\boxed{9}$ $\boxed{.}$ $\boxed{2}$ $\boxed{)}$ $\boxed{\text{ENTER}}$. Your
calculator screen should look like the figure below.

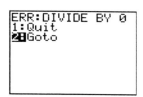

## Division by Zero and Error Messages

Zero cannot be used as a divisor as discussed in Section 1.3. If you divide by zero you will get
an error message on your calculator screen. Try $5 \div 0$ by pressing $\boxed{5}$ $\boxed{\div}$ $\boxed{0}$ $\boxed{\text{ENTER}}$. You will
get the following screen displayed.

```
ERR:DIVIDE BY 0
1:Quit
2■Goto
```

You are given the option to 'Quit' or to 'Goto'. In most cases, you will want to choose 'Goto'
and it will take you to the error. To select 'Goto' you may either press $\boxed{2}$ or you may use the
down arrow key found in the upper right corner and then when the number in front of 'Goto' is
highlighted press $\boxed{\text{ENTER}}$. In this case choosing 'Goto' will place your cursor flashing next to
the zero which was the mistake. You will then be able to edit your entry.

## Exponents and Roots

If you need to square a number enter it and then use the key $\boxed{x^2}$ followed by $\boxed{\text{ENTER}}$. $\boxed{x^2}$ is
located on the right, half way down the key pad. Make sure to use parentheses if the number you
are squaring is negative.

In **Section 1.4, Example 6,** both powers and roots are computed. $(-2)^3$, $\sqrt[3]{-8}$, $3^4$, and $\sqrt[4]{81}$ are discussed. We will compute each of these below.

There are two ways to cube a number. First, begin by entering the number to be cubed. If it contains a minus sign remember to use parenthesis and the $[(-)]$ key. Then we may use the menu option for cubing a number found by pressing the $\boxed{\text{MATH}}$ key. We will select the third item by pressing $\boxed{3}$ to cube a number.

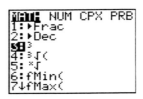

The key strokes you enter for $(-2)^3$ are $\boxed{(}$ $\boxed{(-)}$ $\boxed{2}$ $\boxed{)}$ $\boxed{\text{MATH}}$ $\boxed{3}$ $\boxed{\text{ENTER}}$.

The second way to cube a number is more general and can be used for any exponent. Again you begin by entering the number to be cubed. You then use the carat key $\boxed{\wedge}$ followed by the exponent. The carat key is located on the upper right side of the key pad. $(-2)^3$ is computed by pressing $\boxed{(}$ $\boxed{(-)}$ $\boxed{2}$ $\boxed{)}$ $\boxed{\wedge}$ $\boxed{3}$ $\boxed{\text{ENTER}}$. For a more general exponent such as $3^4$, press $\boxed{3}\boxed{\wedge}\boxed{4}$ $\boxed{\text{ENTER}}$.

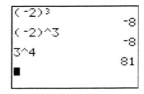

If you need to take the square root of a number press $\boxed{\text{2nd}}$ $[\sqrt{}]$ followed by the number you need to take the square root of. Remember that taking the square root of a negative number will give a non-real answer. For example to get the screen below, type $\boxed{\text{2nd}}$ $[\sqrt{}]$ $\boxed{(-)}$ $\boxed{1}$ $\boxed{)}$ $\boxed{\text{ENTER}}$.

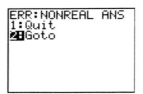

Cubed roots may be found in two ways. The first way is to choose $\boxed{\text{MATH}}$ to get the menu pictured above and select item number 4 which is a raised 3 followed by a square root sign. To find $\sqrt[3]{-8}$ you press, $\boxed{\text{MATH}}$ $\boxed{4}$ $\boxed{(-)}$ $\boxed{8}$ $\boxed{)}$ $\boxed{\text{ENTER}}$. Note that the opening parenthesis automatically will appear. You must type the close parenthesis.

The second way to find a cubed root is more general and can be used for any root. Begin by typing the root you want to take. For example you would type 3 if you wanted the cubed root and 4 if you wanted the fourth root.  Then you go to the $\boxed{\text{MATH}}$ menu and select the fifth item, $\sqrt[x]{\ }$ .  Then enter the number for which you need the root.  Take care to enter parentheses if you have an expression under the root sign.  Parentheses are not automatically entered for the general root.  In the examples below we will not need parentheses since we are using single numbers.

To compute $\sqrt[3]{-8}$ , press $\boxed{3}$ $\boxed{\text{MATH}}$ $\boxed{5}$ $\boxed{(-)}$ $\boxed{8}$ $\boxed{\text{ENTER}}$ .  Notice that this method has the leading 3 large while the first way has a smaller 3.  To compute $\sqrt[4]{81}$ , press $\boxed{4}$ $\boxed{\text{MATH}}$ $\boxed{5}$ $\boxed{8}$ $\boxed{1}$ $\boxed{\text{ENTER}}$ .

```
³√(-8)
                     -2
3 ˣ√-8
                     -2
4 ˣ√81
                      3
■
```

Note that if you take an even root of a negative number, you will get the error message shown earlier for square roots.

## Order of Operations

The TI – 83/83 Plus/84 Plus calculator uses the standard order of operations.  Therefore, as the user it is very important to use parentheses properly.  It is highly recommended that when you enter fractions, enter them using parentheses around the whole fraction.  Further, if the numerator or denominator has more than one number in it make sure to use parentheses around the entire numerator and around the entire denominator.  Finally, if you have a signed number, make sure it is enclosed in parentheses when using exponents.

In **Section 1.5, Example 5**, we are asked to calculate $\dfrac{5}{2-3}+\dfrac{(-4)^2}{-2}$ .  The proper key strokes for this example are $\boxed{(}$ $\boxed{5}$ $\boxed{\div}$ $\boxed{(}$ $\boxed{2}$ $\boxed{-}$ $\boxed{3}$ $\boxed{)}$ $\boxed{)}$ $\boxed{+}$ $\boxed{(}$ $\boxed{(}$ $\boxed{(-)}$ $\boxed{4}$ $\boxed{)}$ $\boxed{x^2}$ $\boxed{\div}$ $\boxed{(-)}$ $\boxed{2}$ $\boxed{)}$ $\boxed{\text{ENTER}}$ .  Note in this example each fraction is enclosed in parenthesis, the denominator (2-3) is in parentheses, and the -4 is in parentheses.  If you get a syntax error, make sure to use 'Goto' by selecting $\boxed{2}$ so that it takes you to the error and you can correct your mistake.

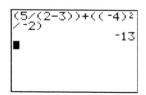

```
(5/(2-3))+((-4)²
/-2)
                   -13
■
```

## Scientific Notation

When doing a computation with very large numbers or very small numbers your calculator may return an answer with an E in it. The E lets you know that scientific notation is being used and the answer must be multiplied by the appropriate power of 10.

In **Section 1.6, Example 9**, the question asks that if one calculation can be done in 0.00000000000026 seconds, how long does it take for 5 million computations? To get the answer we must multiply 0.0000000000026 times 5,000,000.

You may type this directly into your calculator and the answer returned will be 1.3E-6 meaning $1.3 \times 10^{-6}$ or 0.0000013 seconds.

Since it is easy to type the wrong number of zeros into your calculator, it may be better to enter these numbers in scientific notation. To do this, first convert them to scientific notation by hand following the rules in section 1.6 of the text. You get $2.6 \times 10^{-13}$ and $5.0 \times 10^{6}$. To enter a number into the calculator using scientific notation the $\times 10$ is replaced with EE found above the comma key in the center of the keypad and accessed by pressing [2nd] [EE]. To do the computation, press [2] [.] [6] [2nd] [EE] [(-)] [1] [3] [×] [5] [2nd] [EE] [6]. You will get the following screen displayed.

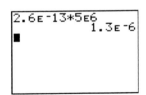

# Chapter 2

# Units of Measurement and Approximate Numbers

## Unit Reduction and Conversion

Chapter 2 of your text book deals with reduction and conversion of units of measure. After multiplying by the required factors for the reduction or conversion, your answer may be very large or very small. If this is the case, watch for E to appear in your answer on the calculator. If it does, your calculator is using scientific notation. For a full discussion of scientific notation on the calculator, refer to Chapter 1.

In **number 3 of the "Now Try It" exercises of section 2.3**, you are asked to convert 30 *ns* to minutes. You set up the expression $30 \, ns \times \dfrac{1 \, s}{10^9 \, ns} \times \dfrac{1 \min}{60 s}$. On your calculator, enter $\boxed{3}$ $\boxed{0}$ $\boxed{\times}$

$\boxed{(}$ $\boxed{1}$ $\boxed{\div}$ $\boxed{1}$ $\boxed{0}$ $\boxed{\wedge}$ $\boxed{9}$ $\boxed{)}$ $\boxed{\times}$ $\boxed{(}$ $\boxed{1}$ $\boxed{\div}$ $\boxed{6}$ $\boxed{0}$ $\boxed{)}$ $\boxed{\text{ENTER}}$. The answer in scientific notation is pictured below. Remember, 5E-1Ø on the calculator represents $5 \times 10^{-10} = 0.0000000005$.

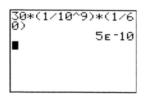

## Rounding Numbers

Frequently we need to round the answers to problems so they have the correct number of significant digits. If the ones place in your answer is significant, then your calculator can help you make sure you round correctly.

In **Section 2.4, Example 8,** you must round your answer to have the correct precision. The numbers 73.2, 8.0627, 93.57 and 66.296 were added to get 241.1287 which with the correct precision should only contain one decimal place.

Press $\boxed{\text{MATH}}$ to access the math menu. Use $\boxed{\blacktriangleright}$ to select NUM sub-menu. Press $\boxed{2}$ to access the round function. Then enter the number $\boxed{2}$ $\boxed{4}$ $\boxed{1}$ $\boxed{.}$ $\boxed{1}$ $\boxed{2}$ $\boxed{8}$ $\boxed{7}$. Follow the number by pressing $\boxed{,}$ $\boxed{1}$ $\boxed{)}$ $\boxed{\text{ENTER}}$. Here the 1 is for one the decimal place required.

In **Example 9,** you divide 292.6 by 3.4 so your answer should have 2 significant digits. When we do the division on the calculator, it displays 86.05882353. In this case, having two significant digits means that we want to round so that we have zero decimal places. This computation can be done by pressing [MATH] [▶] [2] [2] [9] [2] [.] [6] [÷] [3] [.] [4] [,] [0] [)] [ENTER]. Remember that the bolded [MATH] [▶] [2] is to access the round function and the bolded [0] is the number of decimal places desired.

```
round(241.1287,1
)
              241.1
round(292.6/3.4,
0)
                 86
```

In Example 6, you are asked to round 70430 to three significant digits or to 70400. The round function on the calculator will not do this.

# Chapter 3

# Introduction to Algebra

## Entering Pi (π)

**Example 7 in Section 3.1** computes the circumference of the circle using the formula $C = \pi \cdot d$ where $d = 33 \, \text{m}$. The example in the book uses an approximation for $\pi$ but the calculator key 2nd [π] located over the ∧ key on the right of the keypad will give a better approximation for $\pi$. On your calculator press 2nd [π] 3 3 ENTER.

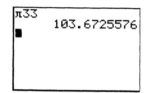

The answer given in the text when 3.14 is multiplied by 33 is 103.62. Clearly the hundredths place does not match but the number of significant digits in the answer is two so the significant digits do match.

# Chapter 4

# Simple Equations and Inequalities

## Checking Solutions

In this chapter you should use your calculator to check your algebraic work. In **Section 4.1, Example 9**, the equation $\frac{x}{3} - \frac{1}{2} = \frac{x}{2} - \frac{5}{2}$ is solved and the answer is given as $x = 12$. Check this on your calculator as two separate problems: $\frac{12}{3} - \frac{1}{2}$ and $\frac{12}{2} - \frac{5}{2}$. Press

─ ( 1 ÷ 2 ) ENTER for the first expression and ( 1 2 ÷ 2 ) ─ ( 5 ÷ 2 ) ) ENTER for the second expression.

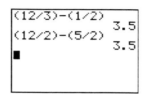

Make sure your check is done in the original equation. While you may be able to do this example in your head, this technique should be used with fractions and/or large numbers.

In Chapter 5, solving simple equations graphically on the calculator is discussed and in Chapter 7 intersection method of solving equations graphically on the calculator is described. These are other ways you will be able to check your work later on in the course.

# Chapter 5

# Graphs

## Introduction

Before beginning it is helpful to find the main keys we will be using on your calculator. [Y=], [WINDOW], [ZOOM], [TRACE] and [GRAPH] are all located just below the calculator screen. Notice that written just above these keys in yellow for the TI-83/83 Plus and blue for the TI-84 Plus are [STAT PLOT], [TBLSET], [FORMAT], [CALC], and [TABLE]. Remember that the [2nd] key must be used before each of these.

The [X,T,Θ,n] is located to the right of the [ALPHA] key and will write the variable X.

In the problem you are given, you may be asked to graph an equation that does not have variables $x$ and $y$ but use different letters. Remember that when graphing or creating a table on the calculator, the independent variable will be entered as $x$ and the dependent variable will be entered as $y$.

## Tables

In **Section 5.1, "Now Try It",** you may create a table of values for the function $f(a) = 5a + 7$ to check your answers for $f(0)$, $f(4)$, $f(-5)$ and $f(12)$. To create the table we will write the function as $y = 5x + 7$.

There are three main steps in creating a table. First, enter the function into the calculator, then set up the desired table parameters and finally display the table.

Begin by entering the function into the calculator. Press the [Y=] button and using the arrow keys, make sure that your cursor is next to \Y₁=. Type [5] [X,T,Θ,n] [+] [7]. Make sure that the equal sign is darkened as pictured below. If the equal sign is not highlighted, using the arrow keys place the cursor over it and press [ENTER].

```
 Plot1 Plot2 Plot3
\Y1■5X+7
\Y2=
\Y3=
\Y4=
\Y5=
\Y6=
\Y7=
```

Next, set the parameters for how we want our table displayed. Press [2nd] [TBLSET]. You use the up and down arrow keys to access the various settings. For this example, set TblStart = 0. This means that $x = 0$ and $y = f(0)$ will be the first row displayed in the table. ΔTbl measures the amount you want the value of *x* to change for each new entry. For this example since we are interested in $x = 0$, 4, -5, and 12 we set $\Delta$Tbl = 1. Indpnt refers to the independent variable *x* and Depend refers to the dependent variable *y*. Choose 'Auto' for this example for both settings. If 'Auto' is not highlighted put your cursor over it using the arrow keys and then press ENTER. We will detail how to use the 'Ask' feature below.

Finally press 2nd [TABLE] to display the table. Notice that all of the desired values do not show. You may use the up and down arrow keys to see more values.

Note that if a value outside your domain is in the X column, the Y₁ column will have the word ERROR.

If the values of *x* that you are interested in are spread out or seem to have no pattern you may go back to the table setup screen ( 2nd [TBLSET] ) and for the independent variable choose 'Ask' by placing your cursor over it and pressing ENTER. Next you press 2nd [TABLE] to display the table and it will be empty with your cursor in the top row. Press the value of *x* that you are interested in followed by ENTER. Your cursor will move down and you will be able to enter more values. You may enter negative number using the (-) key and even do simple mathematical computations.

It is highly recommended that when you are done with the 'Ask' feature of the table setup that you set it back to 'Auto', if you usually want the table to automatically be generated for you.

## Plotting Points

**Section 5.2** discusses plotting points on the rectangular coordinate system. This can be done on the calculator using the [STAT PLOT] feature on the calculator. We will plot the six points found in **Example 2**;

$$P(4, 1), Q(-2, 5), R(-3, -2), S(4, -3), T(-5, 0), \text{ and } V(0, 2).$$

The four steps to plot points are: create the list of points, turn on the plot and select the desired parameters, set the viewing window and display the graph.

To create a list of points choose STAT 1 . The STAT key is located under the DEL key on the top of the keypad. You will then have a table displayed with the first column L1 and the second L2. Place your cursor in the top row of column L1 by using the arrow keys. Your $x$ values of each ordered pair will go in this column. Type 4 ENTER for the first number in the ordered pair P(4, 1). Move your cursor to column L2, row 1 and enter 1 ENTER for the second number $y$ in the ordered pair. Make sure that the $(x, y)$ values that correspond to each other are in the same row or they will not graph properly. Continue in this manner until you have entered all of the points. Remember that negative values must be entered using the (-) key.

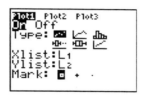

The second step is to turn [STAT PLOT] on. Press 2nd [STAT PLOT]. Select the first plot by pressing 1 . Your cursor will be flashing over ON. Press ENTER to activate the plot. Since we are plotting points we choose the first type of plot, ⊾⁚⁚. Make sure that XList is set to $L_1$ and YList is set to $L_2$. If you need to change these, after highlighting the list you need to change, press CLEAR . Then press 2nd [L1] for list $L_1$ or 2nd [L2] for list $L_2$. You may choose any type of mark you want.

Next we must set our viewing window. To do this we press WINDOW . Move between the settings by using the arrow keys. Since all $x$ values of our ordered pairs are between -6 and 6, these can be used as Xmin and Xmax respectively. Make sure to use (-) when entering the -6. Xscl is the scale that will be used in plotting the $x$ values and is set to 1 for this example. Our $x$-axis will have values from -6 to 6 and each tick mark will represent one unit. Also the $y$ values all fall between -6 and 6 so we may use them for Ymin and Ymax. Yscl can be set to 1 so that the $y$-axis will have values between -6 and 6 and each tick mark represents 1 unit. These window settings are represented

$$X=[-6, 6, 1] \text{ and } Y=[-6, 6, 1].$$

Note that window settings are subjective and as long as you can clearly see the desired data, a different window can be used. Leave Xres = 1. The window setting screen is shown at the top of the next page.

Finally press GRAPH .

 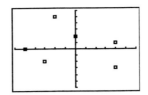

Note that if you have any sort of line across your graph, it may be coming from an equation entered in the [Y=] button. This can be deactivated by pressing [Y=] and then placing your cursor over the equal sign next to any entries that have an equation and press [ENTER]. If the equal sign is not highlighted you have successfully deactivated it. You may need to move the cursor away from the equal sign to see if it is highlighted or not. Of course if you do not plan to use the equation you can clear it. You can go directly back to the graph by pressing [GRAPH].

It is strongly recommended that when you are done with your plot that you turn the plot off. You may modify the directions above for turning the plot on to turn it off. However, the fastest way is to press [Y=], place your cursor over any plot that is highlighted along the top of the screen and press [ENTER]. If you forget to turn a STAT PLOT off, you may get "unexplained" errors when graphing using an equation and different window settings.

## Graphing Functions

Graphing functions has a large overlap with creating tables and stat plots. We will work with **Section 5.3, Example 4** which requires us to graph $y = x^3 - 2$. To graph a function, you enter it into the calculator, set the viewing window and finally display the graph.

As you did for a table, enter the function $x^3 - 2$ into the [Y=] row labeled \Y$_1$. Press [Y=] and then [X,T,$\Theta$,$n$] [^] [3] [−] [2] [ENTER]. Make sure that the equals sign is highlighted. If not cursor over to it using the arrow keys and then press [ENTER].

```
Plot1 Plot2 Plot3
\Y1■X^3-2
\Y2=
\Y3=
\Y4=
\Y5=
\Y6=
\Y7=
```

Next the viewing window must be set and then the function graphed. You may set the window manually or by using one of the calculator's built in settings. The method you use will depend on your objectives.

To manually set the window, press [WINDOW] and enter the following values; Xmin = -6, Xmax = 6, Xscl = 2, Ymin = -80, Ymax = 80, Yscl = 20. These values can be written X= [-6, 6, 2] and Y=[-80, 80, 20]. This will give the window of the graph displayed in the text book. The *x*-axis will go from -6 to 6 with each tick mark representing 2 units. The *y*-axis will go from -80 to 80 with each tick mark representing 20 units.

When setting the window, use the arrow keys to switch between values and remember to use the [(-)] key to enter negative numbers. Press [GRAPH] to draw the graph.

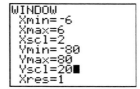  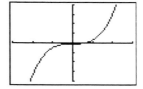

To set the window to the standard viewing window X= [-10, 10, 1] and Y=[-10, 10, 1], you may use [ZOOM] [6]. Note that in this case the graph will automatically be displayed. The resulting graph is pictured below.

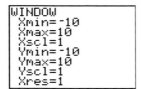  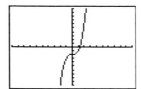

To set the window to "fit" the function, begin by pressing [WINDOW] and entering the desired X values. We will use X=[-6, 6, 1]. You then use [ZOOM] [0]. This will automatically select the Y values for you. You may notice that you can no longer see the vertical tick marks on the *y* - axis as shown in the first graph below. This is because the calculator has not chosen a good scale. Press [WINDOW] and you can see the values for Y. Then choose a better scale for Yscl such as 50 and return to your graph by pressing [GRAPH]. The new window settings and graph are on the right below.

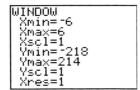  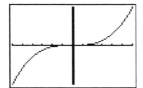  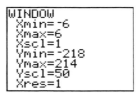  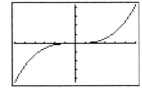

Notice in the three sets of graphs above, that the function looks different depending on the window settings used, but they all graph the same function. You need to determine the window that best works for your application. Use the manual method when you know which part of the graph you need displayed. The standard window can be used if you need a general idea of what the function looks like and you know the graph passes through or near the origin. Often after looking at this view, you will refine your window settings manually. Finally if you are interested in just a certain set of *x* values and want the calculator to determine the range for the *y* values, use [ZOOM] [0].

At any time you may check your window settings by pressing WINDOW and then return directly to the graph by pressing GRAPH. It is important when copying the graph to your handwritten work that you properly label your graph. To determine the increment for the tick marks, refer back to your window settings and look at the scale variables.

Sometimes you will need to know if two lines are perpendicular or if a drawing is a circle or an oval. Because the calculator screen is wider than it is tall, a unit in the $x$ direction is larger than a unit in the $y$ direction. In this case, adjust your window settings using ZOOM 5 which will make the window square. Perpendicular lines will now look perpendicular and circles will look correct.

## Horizontal and Vertical Lines

Horizontal lines have an equation such as $y = 4$ and can be graphed using Y= since they are a function. Clear the equation line that you are going to use. Press 4 ZOOM 6.

Vertical lines have an equation such as $x = 2$ and are not functions so the Y= button may NOT be used. If you want multiple graphs on the same set of axes, enter all of the functions first before using the draw feature discussed below. Make sure you use reasonable settings for your window. The standard window (ZOOM 6) was used in the example below.

To get a vertical line on the calculator you use the [DRAW] feature. **Begin on the main calculator screen.** Press 2nd [DRAW]. Note that [DRAW] is located above PRGM in the upper center of the key board. Choose 4 for Vertical. Then enter the desired value for $x$ which in this example is 2. When you press ENTER, the vertical line will be drawn at $x = 2$.

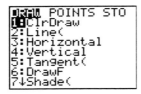 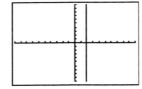

The vertical line will remain each time you press GRAPH, until you edit an equation under Y= and press GRAPH. The vertical lines should then disappear.

## X and Y Intercepts; Finding Zeros

We can use the graphing calculator to find $x$ and $y$ intercepts. In **Section 5.4, Example 4**, we are asked to find the $x$ and $y$ intercepts for the linear function $y = -2x - 5$. Note that if the function was not already solved for $y$, you would need to do the algebra first to solve for $y$.

To find the $x$ and $y$ intercepts, we begin by entering the function using the $\boxed{\text{Y=}}$ button and pressing $\boxed{(\text{-})}$ $\boxed{2}$ $\boxed{\text{X,T,}\Theta,n}$ $\boxed{-}$ $\boxed{5}$. Make sure the equal sign is highlighted. Set your window. For the example below we will use the standard window ($\boxed{\text{ZOOM}}\boxed{6}$).

The $x$ intercepts are where the function crosses the x-axis or where the function value is zero. For the $x$ intercept, press $\boxed{\text{2nd}}$ $\boxed{\text{CALC}}$ and choose option $\boxed{2}$ for zero.

You will see the graph and be prompted for the left bound. Use the arrow keys so that the cursor is to the left of where the graph crosses the $x$-axis and press $\boxed{\text{ENTER}}$. Notice that an arrow is placed on the screen above where your cursor was and you are prompted for the right bound. Move the cursor over so that it is to the right of where the graph crosses the x-axis and press $\boxed{\text{ENTER}}$.

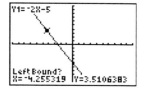
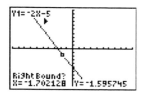

A second arrow is inserted and you are prompted for a guess. You may just press $\boxed{\text{ENTER}}$. The answer will appear at the bottom of the screen as pictured below on the right and is the ordered pair (-2.5, 0).

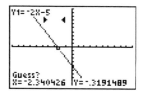
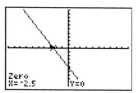

The $y$ intercept represents the value for $y$ when $x$ is zero, so you simply need to evaluate the function for $x = 0$. If you have already entered in the function to find the $x$ intercepts, the quickest way to find the $y$ intercept is to press $\boxed{\text{2nd}}$ $\boxed{\text{CALC}}$ and select $\boxed{1}$ for value. You will be prompted for the $x$ value. Press $\boxed{0}$ and then $\boxed{\text{ENTER}}$. Your answer will appear at the bottom of the screen and in this problem is (0, -5).

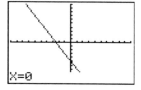
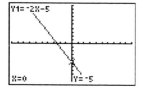

Try to find the $x$ and $y$ intercepts in the equation $y = 2 + \dfrac{1}{x}$ **Section 5.5, Example 2**. When finding the $x$ intercept and setting the right bound, do not cross $x = 0$ where the function is undefined. The calculator will show the error below if you do.

If you try to find the $y$ intercept for this example, the calculator will show the graph but the Y= value will be left blank since the function is not defined at $x = 0$. The graph is pictured on the right below.

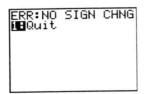     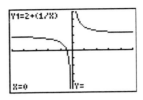

Note that you may use the [TRACE] button to find the coordinates of a point on a graph. Once the graph is displayed you may press [TRACE] and then move the cursor with the arrow keys to the point of interest. Notice that at the bottom of your screen the values of x and y are displayed and as you move they change. This method may not be as accurate as the zero function described above.

## Solving Equations Graphically Using Zeros

This is the same as finding the $x$ intercept if the equation is set equal to zero. Refer to the previous section for an introduction on $x$ and $y$ intercepts.

In **Section 5.6, Example 5**, you are asked to solve $x^2 - 4x + 2 = 0$ graphically.

Begin entering the equation by pressing [Y=] and then [X,T,Θ,n] [x²] [−] [4] [X,T,Θ,n] [+] [2]. Next set your window by pressing [WINDOW]. We will use the settings X=[-2, 6, 1] and Y=[-3, 7, 1]. For a full discussion on window settings see the section on graphing functions. Finally, press [GRAPH].

     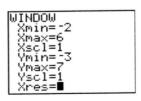     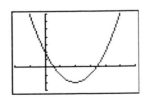

Now you are set up to solve the equation or find the zeros. First press [2nd] [CALC] and then choose [2] for zero. You will be prompted for the left bound. Using the arrow keys move to the left of the zero that is approximately 0.5 and press [ENTER]. You will then be prompted for the right bound. Using the arrow keys move to the right of the zero near 0.5 but do NOT cross the next zero near 3.5. (See the middle picture on the top of the next page.) Press [ENTER]. Finally, you will be prompted for a guess. You may just press ENTER again. The bottom of the calculator screen (pictured on the left below) shows that there is a zero for x ≈ 0.58578644.

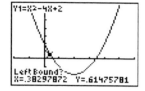 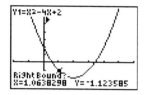 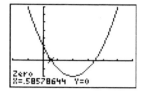

You must repeat this procedure for the second zero near 3.5. Here when setting the left bound make sure the cursor is to the right of the zero we just found or the calculator may find the zero we already know about. The final screen should show that there is a zero for x ≈ 3.4142136. Pictured below are the left bound screen and the final screen.

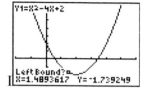 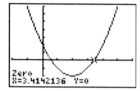

To solve more general equations such as $x^2 + 5x = 3x + 2$ you may view these as two functions $Y_1 = x^2 + 5x$ and $Y_2 = 3x + 2$ and the solution will be at the intersection of the graphs or where they are equal. You learned to solve this algebraically in Chapter 3 of your book and will learn about using the calculator to find intersections in Chapter 7 of this manual. Chapter 9 of this manual demonstrates a method for solving the general equation in one variable.

## Maximum and Minimum Values Graphically

Review the section on graphing for entering a function into the calculator and for setting the window parameters.

In **Section 5.6, Example 4**, you are given the height of an object above the ground to be $d = -16t^2 + 128t$ and among other things are asked to find the maximum height of the object. The independent variable is $t$ and the dependent variable is $d$ so we may re-write the equation as $y = -16x^2 + 128x$ where $x$ now represents the time and $y$ represents the height of the object.

Enter $y = -16x^2 + 128x$ for \Y₁ in Y= screen. Using WINDOW, enter X=[0, 10, 1] and Y=[0, 300, 10]. Note that the minimum values for $x$ and $y$ were set to zero because negative time does not make sense and the problem assumes the ground is level so you can't get a height below zero. The maximum value for $y$ was chosen after ZOOM 0 was used.

We are now ready to find the maximum value. Use 2nd [CALC] and choose 4 for maximum. You will be prompted for the left bound. Cursor to the left of where you think the maximum is and press ENTER. You will then be prompted for the right bound so you cursor to the right of the maximum and press ENTER. Finally you will be prompted for a guess. Notice in the picture below on the right that there are arrows where the left and right bounds were marked. When asked for the guess, you may just press ENTER. Your answer will be displayed at the bottom of the screen and is at $x = 3.9999987$ seconds with a height of $y = 256$ feet. Using 2nd [CALC] 1 and entering $x = 4$ seconds gives a height of 256 feet so $x$ should be rounded to 4 for the final answer.

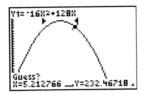 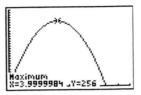

If you require a minimum, the same procedure is used however after pressing 2nd [CALC] choose 3 for minimum. The remainder of the procedure is the same.

## Graphing Inequalities

Graphing inequality is very similar to graphing a function. Refer to the section on graphing functions for a refresher. When given an inequality, you must first solve for the dependent variable, usually $y$.

In **Section 5.7, Example 5**, the inequality $y \geq x^2$ is graphed.

When graphing by hand you would begin by graphing $y = x^2$. Enter this equality for \Y$_1$ when using the Y= button. Before leaving this screen, use the arrow keys to cursor over to the left of Y$_1$. Press ENTER until you see the symbol ▜. We choose ▜ because $y$ is greater than or equal to $x^2$. Set your window and then display your graph. In the following, graph the window settings are X=[-4, 4, 1] and Y=[-1, 10, 1].

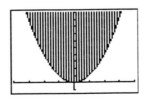

If you are graphing an inequality where $y$ is less than or equal to an equation in $x$ you would choose the ▙ symbol.

# Chapter 6

# Introduction to Geometry

## Degrees: As Decimals and as Degrees Minutes Seconds (DMS)

Chapter 6 introduces angles as decimals and in the degree minute second form (DMS). The TI-83/83 Plus/84 Plus easily converts degrees from one form to the other. We will use the ANGLE menu found by pressing 2nd [ANGLE].

**Section 6.1, Example 2** gives 17.2° and asks for it to be converted to DMS form. On your calculator, press 1 7 . 2 2nd [ANGLE] 4 ENTER. The calculator returns 17°12´.

The second part of the question asks for 58°36´ to be converted to a decimal. To do this, press 5 8 2nd [ANGLE] 1 3 6 2nd [ANGLE] 2 ENTER. The calculator returns 58.6. Note that this measurement is in degrees even though the calculator does not indicate it. 2nd [ANGLE] 1 is used to enter degrees and 2nd [ANGLE] 2 is used to enter the minutes.

```
17.2▶DMS
           17°12'0"
58°36'
              58.6
```

## Pi in Geometry

In Chapter 3, we saw $\pi$ was used in computing the circumference of a circle. The area of a circle and the volume of a sphere also use $\pi$ in their formulas. When using these formulas make sure that your exponent on the calculator only applies to the radius. You will not need parenthesis for these two formulas.

In **Example 6 from Section 6.3**, the area of a circle of radius 2.73 cm is $A = \pi(2.73cm)^2$. Enter 2nd [$\pi$] 2 . 7 3 $x^2$ ENTER. This gives 23.41397589. Since this example has 3 significant digits, we need to keep one decimal place. Using the round feature from Chapter 2, enter MATH ▶ 2 2nd [ANS] , 1 ) ENTER . Note that 2nd [ANS] recalls the previous answer found. The final answer is 23.4.

```
π2.73²
            23.41397589
round(Ans,1)
                 23.4
```

# Chapter 7

# Simultaneous Equations

## Solve a System of Linear Equations Graphically; Intersections

In **Section 7.1, Example 4** we are asked to solve the following system of equations graphically;

$$x + y = 9$$
$$x - y = 2$$

The first step is to re-write them so that they are in $y = mx + b$ form

$$y = -x + 9$$
$$y = x - 2$$

Using $\boxed{Y=}$, enter in $Y_1 = -x + 9$ and $Y_2 = x - 2$. Display the graph in the standard window by using $\boxed{ZOOM}$ $\boxed{6}$.

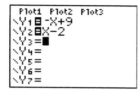 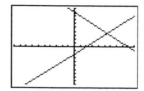

Then press $\boxed{2nd}$ $\boxed{CALC}$ and choose $\boxed{5}$ for intersection. You will be prompted to choose the first curve. Use the arrow keys to cursor over to the intersection of choice and choose $\boxed{ENTER}$. You then will be prompted for the second curve. If the calculator does not move your cursor to the second curve for you, use either the up or down arrow key to switch curves and again press $\boxed{ENTER}$. You will then be prompted for a guess (not pictured). It is recommended that you put the cursor near the intersection before pressing $\boxed{ENTER}$. Your result will be displayed. The intersection is (5.5, 3.5) meaning that $x = 5.5$ and $y = 3.5$ is a solution to the original system of equations.

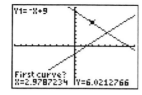 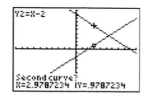 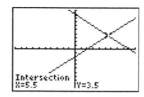

Always check your answers in the original equations to make sure that you did not make an algebraic mistake preparing for the problem setup.

## Addition-Subtraction Method; rref()

The calculator uses a similar method to addition-subtraction for solving systems of linear equations that are in standard form, $ax + by = c$. If your equations are not in this form, you must do the necessary algebra to get them in the correct form first.

**Example 3 of Section 7.3** asks for the following system of equations to be solved.

$$4x - y = 4$$
$$2x + 3y = -5$$

Since these equations are in standard form, we may begin entering them into a matrix. Press MATRX on the TI-83 or 2nd [MATRIX] on the TI-83 Plus/84 Plus to get the matrix menu with headings NAMES, MATH, EDIT.

Move the cursor over to EDIT by pressing ▶ ▶ and then press 1 for matrix [A].

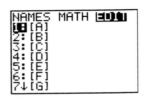

Next you will enter the size which is 2 by 3. Press 2 ▶ 3 ▶ and at this point you should see 2 rows each filled with 3 numbers. Your cursor should be highlighting the number in the first row and the first column.

The coefficients from the first equation go in the first row of the matrix and those from the second equation go in the second row of the matrix. The coefficient for $x$ must go in the first column and the coefficient for $y$ must go in the second column. Input each number followed by the ENTER key to move to the next location. Remember to use (-) for negative signed numbers and that the coefficient in front of $y$ in the second equation should be viewed as plus -1. If you make a mistake entering any of the coefficients you may use the arrow keys to go back and correct it. Your screen should match the one pictured below.

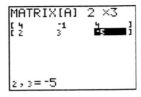

Press 2nd [QUIT] to leave the matrix entry screen.

Next for the TI-83, press MATRX ▶ or for the TI-83 Plus/84 Plus, press 2nd MATRIX ▶ to access the operations on the matrix MATH menu that can be performed on a matrix. For either calculator, you will choose rref (either by pressing ALPHA [B] or by using the down arrow key until it is highlighted and then pressing ENTER. Next to access matrix [A], press MATRX 1 for the TI-83 or 2nd MATRIX 1 for the TI-83 Plus/84 Plus. For either calculator, close the parenthesis, ), and press ENTER. Your screen will look like the picture below.

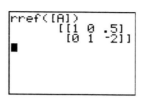

Your answer appears in the last column and is $x = .5$ and $y = -2$. Remember each row represents an equation with the $x$ coefficient in the first column and the $y$ coefficient in the second column. Therefore the first row is read $1x + 0y = .5$ or simplified $x = .5$. Similarly, the second row is interpreted as $0x + 1y = -2$ or simplified $y = -2$.

## Determinants

Using the method of determinants, in **Example 4 of Section 7.4**, you need to solve

$$1.34x - 2.73y = 9.44$$
$$-8.35 + 7.22y = 5.36$$

First enter the three required 2 by 2 matrices;

$$A = \begin{vmatrix} 1.34 & -2.73 \\ -8.35 & 7.22 \end{vmatrix}, B = \begin{vmatrix} 9.44 & -2.73 \\ 5.36 & 7.22 \end{vmatrix}, C = \begin{vmatrix} 1.34 & 9.44 \\ -8.35 & 5.36 \end{vmatrix}$$

For details on entering a matrix, see the previous section.

For this example, you use the following key strokes for matrix [A]; For the TI-83 press MATRX or for the TI-83 Plus/84 Plus press 2nd[MATRIX]. Then for any of the three calculators press ▶ ▶ ENTER 2 ▶ 2 ▶ 1 . 3 4 ENTER (-) 2 . 7 3 ENTER (-) 8 . 3 5 ENTER 7 . 2 2 ENTER 2nd[QUIT].

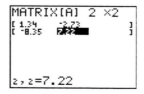

For matrix $[B]$, on the TI-83 press $\boxed{\text{MATRX}}$ or on the TI-83 Plus/84 Plus press $\boxed{\text{2nd}}\boxed{\text{MATRIX}}$. For any of the three calculators press $\boxed{\blacktriangleright}\boxed{\blacktriangleright}\boxed{2}$ $\boxed{2}\boxed{\blacktriangleright}\boxed{2}\boxed{\blacktriangleright}$ $\boxed{9}\boxed{.}\boxed{4}\boxed{4}\boxed{\text{ENTER}}$ $\boxed{(\text{-})}\boxed{2}\boxed{.}\boxed{7}\boxed{3}\boxed{\text{ENTER}}$ $\boxed{5}\boxed{.}\boxed{3}\boxed{6}\boxed{\text{ENTER}}$ $\boxed{7}\boxed{.}\boxed{2}\boxed{2}\boxed{\text{ENTER}}$ $\boxed{\text{2nd}}\boxed{\text{QUIT}}$.

For matrix $[C]$, on the TI-83 press $\boxed{\text{MATRX}}$ or on the TI-83 Plus/84 Plus press $\boxed{\text{2nd}}\boxed{\text{MATRIX}}$. Then for any of the three calculators, press $\boxed{\blacktriangleright}\boxed{\blacktriangleright}\boxed{3}$ $\boxed{2}\boxed{\blacktriangleright}\boxed{2}\boxed{\blacktriangleright}$ $\boxed{1}\boxed{.}\boxed{3}\boxed{4}\boxed{\text{ENTER}}$ $\boxed{9}\boxed{.}\boxed{4}\boxed{4}\boxed{\text{ENTER}}$ $\boxed{(\text{-})}\boxed{8}\boxed{.}\boxed{3}\boxed{5}\boxed{\text{ENTER}}$ $\boxed{5}\boxed{.}\boxed{3}\boxed{6}\boxed{\text{ENTER}}$ $\boxed{\text{2nd}}\boxed{\text{QUIT}}$.

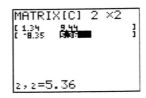

To find the value of $x$, we calculate $\dfrac{\det([B])}{\det([A])}$.

On the TI-83 press $\boxed{\text{MATRX}}$ $\boxed{\blacktriangleright}$ to access the MATRIX menu MATH functions. Press $\boxed{1}$ for det ( for the determinant. Press $\boxed{\text{MATRX}}$ $\boxed{2}$ to choose matrix $[B]$. Press $\boxed{)}$ $\boxed{\div}$. Again, press Press $\boxed{\text{MATRX}}$ $\boxed{\blacktriangleright}$ $\boxed{1}$ for det ( for the determinant. Press $\boxed{\text{MATRX}}$ $\boxed{1}$ to choose matrix $[A]$. Finally press $\boxed{)}$ $\boxed{\text{ENTER}}$.

On the TI-83 Plus/84 Plus press $\boxed{\text{2nd}}$ $\boxed{\text{MATRIX}}$ $\boxed{\blacktriangleright}$ to access the MATRIX menu MATH functions. Press $\boxed{1}$ for det ( for the determinant. Press $\boxed{\text{2nd}}$ $\boxed{\text{MATRIX}}$ $\boxed{2}$ to choose matrix $[B]$. Press $\boxed{)}$ $\boxed{\div}$. Again, press $\boxed{\text{2nd}}$ $\boxed{\text{MATRIX}}$ $\boxed{\blacktriangleright}$ $\boxed{1}$ for det ( for the determinant. Press $\boxed{\text{2nd}}$ $\boxed{\text{MATRIX}}$ $\boxed{1}$ to choose matrix $[A]$. Finally press $\boxed{)}$ $\boxed{\text{ENTER}}$.

The answer is given as -6.309846273. We want three significant digits so this is rounded to 2 decimal places by pressing $\boxed{\text{MATH}}$ $\boxed{\blacktriangleright}$ $\boxed{2}$ $\boxed{\text{2nd}}$ $\boxed{\text{ANS}}$ $\boxed{,}$ $\boxed{2}$ $\boxed{)}$ $\boxed{\text{ENTER}}$ (see Chapter 2).

```
det([B])/det([A]
)
        -6.309846273
round(Ans,2)
               -6.31
```

To find the value of *y*, we calculate $\dfrac{\det([C])}{\det([A])}$ .

For the TI-83 the key strokes are [MATRX] [▶] [1] [MATRX] [3] [)] [÷] [MATRX] [▶] [1] [MATRX] [1] [)] [ENTER].  Note that the only difference in the key strokes between the *x* and *y* values is the bolded [3].

For the TI-83 Plus/84 Plus the key strokes are [2nd] [MATRIX] [▶] [1] [2nd] [MATRIX] [3] [)] [÷] [2nd] [MATRIX] [▶] [1] [2nd] [MATRIX] [1] [)] [ENTER].  Note that the only difference in the key strokes between the *x* and *y* values is the bolded [3].

The answer is given as -6.55501612.  We want three significant digits so this is rounded to 2 decimal places by pressing [MATH] [▶] [2] [2nd] [ANS] [,] [2] [)] [ENTER].

```
det([C])/det([A]
)
         -6.55501612
round(Ans,2)
               -6.56
```

# Chapter 8

# Factoring

## Factoring Graphically

The calculator can be helpful in factoring more difficult problems. For example, in **Exercise 25 of Section 8.2** you need to factor $x^2 - 3x - 108$. Because 108 has many factors, finding the zeros of the equation can help us to make the correct choice of factors.

Begin by entering the equation $x^2 - 3x - 108$ in the $\boxed{Y=}$ screen (see Graphing Functions / Finding Zeros in Chapter 5 for a refresher). If you use the standard window, you will only see a small piece of the function, however it is sufficient because we can see one location where the graph crosses the $x$ axis. Find this zero using $\boxed{\text{2nd}}$ $\boxed{\text{CALC}}$ $\boxed{2}$. Here when prompted for the left bound you may want to enter $\boxed{(-)}$ $\boxed{1}$ $\boxed{0}$ $\boxed{\text{ENTER}}$ rather than using the arrow keys. Then for the right bound enter $\boxed{(-)}$ $\boxed{8}$ $\boxed{\text{ENTER}}$ instead of using the arrow keys. Finally press $\boxed{\text{ENTER}}$ and you should find that $x = -9$ is a zero. This suggests that $x + 9$ is a factor. Notice the sign change when guessing at the factor. The reason for this guess can be derived from Chapter 11, Section 2 on solving quadratic equations by factoring. Once you know one factor, you can determine the other. Because two numbers must multiply to -108 and one is 9 the other is -12. The sum of these is -3 and our factors are $(x + 9)(x - 12)$. Check your answer by multiplying.

Another example where the calculator is helpful is **"Now Try It!" Exercise 3 of Section 8.3**. Here $6x^2 + 7x - 20$ needs to be factored. Find the zero on the left using your calculator. The answer is $x = -2.5 = \dfrac{-5}{2}$ suggesting that $2x + 5$ may be a factor. Notice here that the denominator was placed in front of the $x$ and the numerator of the fraction was used as the number added with a sign change. If $2x + 5$ is one factor, the other is $3x - 4$ because $2 \times 3 = 6$ and $5 \times -4 = -20$. The answer is $(2x + 5)(3x - 4)$.

To find a fraction from a decimal, enter the number into the calculator and then use $\boxed{\text{MATH}}$ $\boxed{1}$ $\boxed{\text{ENTER}}$. If we had chosen the right hand zero in the example above, our zero would have been 1.3333333. If you try to convert this to a fraction the calculator will just return a decimal. After noticing that the three repeats we can enter 1.333333333333 (3 appears here 11 times) and then convert using $\boxed{\text{MATH}}$ $\boxed{1}$ $\boxed{\text{ENTER}}$. The calculator returns 4/3.

In cases where there is an integer zero, make sure to use that one. Not all equations factor as nicely as these do. If there is not an integral or fractional answer, this method will not work for factoring.

# Chapter 9

# Algebraic Functions

## Check Your Algebraic Work

When algebraic expressions are simplified, you may use your calculator to check your work. In **Section 9.1, Example 4**, the expression $\dfrac{x^2 - x - 2}{x^2 + 3x + 2}$ is simplified to $\dfrac{x - 2}{x + 2}$. Each of these expressions needs to be entered into the Y= screen as pictured below. Make sure to use parentheses around both the numerators and the denominators. For the first equation, move the cursor in front of the Y₁= and press ENTER until you see the dotted line, ⋅⋅. Notice that the dotted line has more space between the dots than the standard line. Each time the equation is edited or a new equation is entered this setting returns to the standard line.

```
Plot1 Plot2 Plot3
·.\Y1B(X²-X-2)/(X²
+3X+2)
\Y2B(X-2)/(X+2)
\Y3=■
\Y4=
\Y5=
\Y6=
```

Set your window and graph the equations. When the lines for the graph are drawn, the dotted line will be drawn first so you can see the second one drawn over it. If the graphs are not on top of each other there is an error in your calculations. Be careful. The graphs may look like they are on top of each other when they are not. Check the table of values as well. To match the table below remember to use 2nd [TABLE] and set the values as pictured below.

Here the $Y_1$ and $Y_2$ values in each row of the table match except for $x = -1$. In this case the first equation has an undefined denominator when the second is not. This is the exception. If this is the only place that the tables do not match, you have done your algebra correctly.

To see an example where the graphs appear the same but your work is incorrect, edit the example above so that the first expression is $\dfrac{x^2 - x - 1}{x^2 + 3x + 2}$. The only thing you need to change is the 2 in the numerator of the first fraction. The graphs still appear to be on top of each other. Look at the table. The values for $Y_1$ and $Y_2$ in each row are different, indicating that a mistake was made.

## Solve Equations Graphically Using Intersections

The TI-83/83 Plus/84 Plus can solve equations with one variable graphically. We will solve **Example 4 of Section 9.5**; $\dfrac{2}{t} = \dfrac{3}{t+2}$. We begin viewing this as two separate equations; $Y_1 = 2/x$ and $Y_2 = 3/(x+2)$. Note that here $t$ is the independent variable so it may be replaced with $x$ when using the calculator. Enter these two equations using the $\boxed{Y=}$ button. Make sure to use parenthesis around the second denominator.

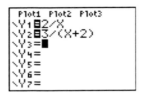

Next you must choose your window. The graph pictured below uses the standard window ($\boxed{ZOOM}$ $\boxed{6}$). The window must be set so that the intersections are identifiable. While we can't see the intersection clearly, it is likely the intersection is above the positive $x$ axis and is in our window. Press $\boxed{2nd}$ $\boxed{CALC}$ $\boxed{5}$. Make sure the cursor is on a curve above the positive $x$ axis (not $x = 0$) and then press $\boxed{ENTER}$ $\boxed{ENTER}$ $\boxed{ENTER}$. The intersection is (4, 0.5).

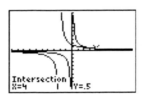

This equation only has one intersection, but watch for those that have more than one, especially those that have intersections that don't appear in the standard window.

# Chapter 10

# Exponents, Roots and Radicals

There are many simplifications and equations solved in this chapter algebraically.  Refer to Chapter 9 of this manual to review how you can use the calculator to check your work.

Note that simplifications are not approximations.  Problems such as $\sqrt{72}$ cannot be simplified on the calculator, only approximated.  However the approximation can be useful.  For example, check the approximation of $\sqrt{72}$ with the approximation of the simplification $6\sqrt{2}$ and make sure they are the same.

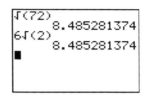

## Entering Fractional Exponents

Integer exponents were covered in Chapter 1 of this manual.

In **Section 10.2, Example 4**, you are asked to simplify $8^{2/3}$.  You can re-write this as a radical and use the techniques of Chapter 1 to enter the problem into the calculator.  Alternatively, you may use 2/3 as the exponent making sure that you use parenthesis around the 2/3.  Press $\boxed{8}$ $\boxed{\wedge}$ $\boxed{(}$ $\boxed{2}$ $\boxed{\div}$ $\boxed{3}$ $\boxed{)}$ $\boxed{\text{ENTER}}$.

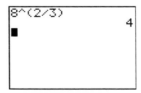

## Complex Numbers

Typically if you enter a negative square root into your calculator you get an error as discussed in Chapter 1 of this manual.  A negative square root returns a complex answer.

For your calculator to give you the complex answer, you must adjust the mode settings.  Press
MODE .  In the row starting with REAL, select  $a + bi$  by placing the cursor over it and by
pressing ENTER .  Your screen should look like the picture below.

Use 2nd [QUIT] to leave the mode screen.

Now enter  $\sqrt{-16}$  from **Example 2 of Section 10.3**.  Press 2nd [√] (-) 1 6 ) ENTER .  The

calculator returns  $4i$ .  Remember that we are using (-) to enter the negative signed number.
Note that the calculator uses the letter  $i$  to represent an imaginary number while the book uses
the letter  $j$ .  You no longer get an error message because of the mode setting.

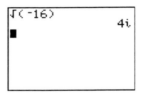

# Chapter 11

# Quadratic Equations

## Complex Numbers as Solutions

In **Section 11.1, Example 5**, we are given that *2i* and *-2i* are solutions to $x^2 + 4 = 0$. (Recall that the book uses the letter *j* to represent imaginary numbers but the calculator uses *i*.) You may verify that *2i* and *-2i* are solutions to $x^2 + 4 = 0$ on your calculator by using the $[i]$ key. Press $\boxed{(}$ $\boxed{2}$ $\boxed{\text{2nd}}$ $[i]$ $\boxed{)}$ $\boxed{x^2}$ $\boxed{+}$ $\boxed{4}$ $\boxed{\text{ENTER}}$ for the first solution, *2i*. For the second solution, *-2i*, type $\boxed{(}$ $\boxed{(-)}$ $\boxed{2}$ $\boxed{\text{2nd}}$ $[i]$ $\boxed{)}$ $\boxed{x^2}$ $\boxed{+}$ $\boxed{4}$ $\boxed{\text{ENTER}}$. In each case, we get the answer of zero returned indicating our solution was correct.

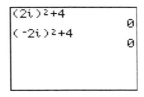

Note that here you do not need to adjust the mode settings. Either real or complex can be selected.

## Factoring Graphically Revisited

In Chapter 8, we used the graph of a function to help determine the factors of an algebraic expression. After obtaining the graph, we found the zeros and were able to find the factor based on the values of the zeros.

In **Section 11.2, Example 3**, we must solve $6x^2 + 7x - 5 = 0$ by factoring. Suppose you are having trouble finding the correct factors for $6x^2 + 7x - 5$. Enter the expression into the $\boxed{Y=}$ screen, set the viewing window and graph the function. Use $\boxed{\text{2nd}}$ $\boxed{\text{CALC}}$ $\boxed{2}$ to find the zeros. For review on finding a zero, you can revisit Chapter 5 of this manual. The zeros are $x = \frac{1}{2}$ and $x = -\frac{5}{3}$. Our factors would then be $(2x - 1)(3x + 5)$. Remember that to determine the factors the denominator of the solution was placed in front of the *x* and the numerator was the number added with a sign change.

We will finish the problem algebraically to show why the choice for the factors is justified. We now have that $(2x - 1)(3x + 5) = 0$. Apply the zero property to get the following two equations: $2x - 1 = 0$ and $3x + 5 = 0$. Solve. The first equation gives $x = \frac{1}{2}$ and the second gives $x = -\frac{5}{3}$. These were the zeros we found graphically. In a sense the graphing calculator allowed us to jump to the answer and then work backward to factor.

## The Quadratic Formula

To solve the equation $ax^2 + bx + c = 0$, you use the formula $x = \dfrac{-b \pm \sqrt{b^2 - 4ac}}{2a}$. **In Section**

**11.4, Example 1**, we are asked to solve $x^2 - 5x + 6 = 0$. Pick out $a$, $b$, and $c$. $a = 1$, $b = -5$ and $c = 6$. Substitute these into the formula above to get the equation

$x = \dfrac{-(-5) \pm \sqrt{(-5)^2 - 4 \times 1 \times 6}}{2(1)}$. Notice the careful use of parenthesis especially around the

negative 5 that was plugged in for $b^2$. You must include these parentheses when using your calculator. Also the numerator and denominator each need to be written with parenthesis around them.

Because the equation uses the symbol $\pm$ read "plus or minus" two computations must be done; one with the + sign and one with the − sign. Since each computation will require the

discriminant $D = \sqrt{(-5)^2 - 4 \times 1 \times 6}$ to be computed, we will store this value in our calculator

first. Press $\boxed{\text{2nd}}$ $\boxed{\sqrt{\;}}$ $\boxed{(}$ $\boxed{(\text{-})}$ $\boxed{5}$ $\boxed{)}$ $\boxed{x^2}$ $\boxed{-}$ $\boxed{4}$ $\boxed{\times}$ $\boxed{1}$ $\boxed{\times}$ $\boxed{6}$ $\boxed{)}$ $\boxed{\text{STO}\blacktriangleright}$ $\boxed{\text{ALPHA}}$ $[\,\text{D}\,]$ $\boxed{\text{ENTER}}$.

(Note that $\boxed{\text{ALPHA}}$ key is green so to find $[\,\text{D}\,]$ you are looking at the green lettering for D.) This does the computation and stores the value into D. You can then access that value by

pressing $\boxed{\text{ALPHA}}$ $[\,\text{D}\,]$ $\boxed{\text{ENTER}}$ at any time.

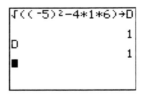

We now need to calculate $\dfrac{-(-5) + D}{2 \times 1}$. To do this, press $\boxed{(}$ $\boxed{(\text{-})}$ $\boxed{(}$ $\boxed{(\text{-})}$ $\boxed{5}$ $\boxed{)}$ $\boxed{+}$ $\boxed{\text{ALPHA}}$ $[\,\text{D}\,]$

$\boxed{)}$ $\boxed{\div}$ $\boxed{(}$ $\boxed{2}$ $\boxed{\times}$ $\boxed{1}$ $\boxed{)}$ $\boxed{\text{ENTER}}$. We get the answer $x = 3$.

Next we calculate $\dfrac{-(-5) - D}{2 \times 1}$. Rather than retyping the last entry and just changing the plus to a

minus, press $\boxed{\text{2nd}}$ $\boxed{\text{ENTER}}$ and it will recall your previous computation. Use the $\boxed{\blacktriangleleft}$ key until

your cursor is over the plus sign and then press $\boxed{-}$ to replace the plus and $\boxed{\text{ENTER}}$ for the computation to occur. The second answer is $x = 2$.

```
(-(-5)+D)/(2*1)
                3
(-(-5)-D)/(2*1)
                2
```

## Finding the Vertex; Maximums and Minimums

The vertex of a quadratic function is a maximum when the coefficient in front of the $x^2$ is negative and a minimum when the coefficient in front of the $x^2$ is positive. In Chapter 5 we discussed finding the maximum and minimum of a function.

In **Section 11.5, Example 4** we are asked to find the vertex of $y = -8x^2 + 24x - 3$. Since a = -8 we know that the graph opens down and we must have a maximum. Enter the equation $y = -8x^2 + 24x - 3$ into the [Y=] screen. Set your viewing window so that you can clearly see the maximum. Use [ZOOM] [6] to start with the standard window. The maximum is not visible, but you can tell that it is between $x = 0$ and $x = 3$. Adjust the $x$ values using [WINDOW] to X=[-1, 4, 1] and then use [ZOOM] [0] for 'ZoomFit'. You can now see the maximum.

Press [2nd] [CALC] [4] for maximum. You can enter the left and right bounds and your guess using the arrow and enter keys as in Chapter 5 or you may actually type values. When prompted for the left bound type [1] [ENTER]. Notice that when you press a number the bottom of the screen changes so it reads X=1 before you press enter. We use 1 because it is clearly to the left of the maximum.

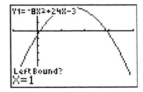

When prompted for the right bound use [2] [ENTER] because 2 is clearly to the right of the maximum. Here the bottom of the screen shows X=2 before you press enter.

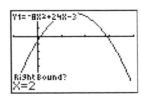

You may then just press [ENTER] for the guess. The calculator will display both the $x$ and $y$ values which make up the vertex, (1.5, 15).

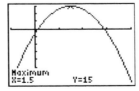

## Solve a Quadratic Equation Graphically

In **Section 11.5, Example 5** we are asked to find the roots of $y = x^2 - x - 6$. The roots are the $x$-intercepts and can be found graphically on the calculator by using the zero feature. Refer to the section on $x$-intercepts in Chapter 5 for complete details on finding zeros graphically.

Enter $y = x^2 - x - 6$ in the $\boxed{Y=}$ screen. Set your window values using $\boxed{ZOOM}$ $\boxed{6}$. Two $x$-intercepts will be visible. Press $\boxed{2nd}$ $\boxed{CALC}$ $\boxed{2}$ to begin finding the left intercept. When prompted for the left bound you can type a value such as $\boxed{(-)}$ $\boxed{3}$ followed by $\boxed{ENTER}$ rather than using the arrow keys. Note, any value to the left of the intercept could be entered, but a value reasonably close should be used. In this case -3 works well. After you entered the value -3 before pressing enter, notice that the bottom left corner is displaying the value for $x$. When prompted for the right bound type $\boxed{(-)}$ $\boxed{1}$ $\boxed{ENTER}$. Again we need a value that is to the right of the zero but be careful not to choose a value greater than the right hand zero. We chose -1 because it was to the right of the left zero but reasonably close. For the guess, just press $\boxed{ENTER}$. This gives the first zero (-2, 0).

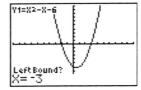 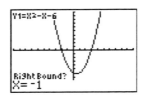 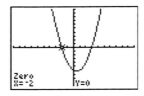

Find the right intercept by again pressing $\boxed{2nd}$ $\boxed{CALC}$ $\boxed{2}$. For the left and right bounds you can use the values 2 and 4 respectively. Press $\boxed{2}$ $\boxed{ENTER}$ $\boxed{4}$ $\boxed{ENTER}$ $\boxed{ENTER}$. The zero (3, 0) will be displayed.

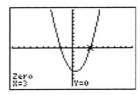

The roots to the equation are $x = $ -2 and 3.

# Chapter 12

# Exponential and Logarithmic Functions

## Evaluating Exponential and Logarithmic Functions

Integer exponent evaluation was done in Chapter 1 and fractional exponents were covered in Chapter 10.

The constant $e$ often serves as the base in exponential functions and has its own key on the calculator. It is located two keys above the $\boxed{\text{ON}}$ key on the left side of the keypad. To access it you press $\boxed{\text{2nd}}$ $[e^x]$. Notice that the exponent symbol, ^, and the opening parenthesis are automatically entered for you.

For example if you wanted to evaluate $5e^{3.4}$ you would press $\boxed{5}$ $\boxed{\text{2nd}}$ $[e^x]$ $\boxed{3}$ $\boxed{.}$ $\boxed{4}$ $\boxed{)}$ $\boxed{\text{ENTER}}$. Your screen should look like the one pictured below.

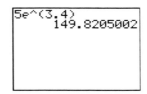

In **Example 3a of Section 12.2**, we are asked to evaluate $\log(58.24)$. To access the logarithm function, press the $\boxed{\text{LOG}}$ key on the left side of the keypad. Notice that it already opens the parenthesis for you. Simply type the number, close the parenthesis and evaluate.

Your key presses are $\boxed{\text{LOG}}$ $\boxed{5}$ $\boxed{8}$ $\boxed{.}$ $\boxed{2}$ $\boxed{4}$ $\boxed{)}$ $\boxed{\text{ENTER}}$.

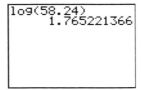

The domain for the logarithm is all positive real numbers. You will receive the error message on the left if you try to evaluate $\log(0)$ and the one on the right if you enter a negative number.

 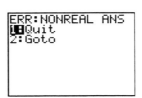

The natural logarithm is evaluated in a similar manner to log base 10 except you use the $\boxed{\text{LN}}$ key found just below the $\boxed{\text{LOG}}$ key.

In **Section 12.4, Example 4** asks for $\ln(1.5)^2$ to be evaluated. You may use the properties to simplify $\ln(1.5)^2$ as the book does, or enter it directly into the calculator. When entering $\ln(1.5)^2$ into your calculator you must be careful about your parentheses. If you type exactly what you see the calculator will take the natural logarithm of 1.5 and square the answer. This is not what we want. We want 1.5 squared and then have the natural logarithm taken. Parentheses should be added around the 1.5 squared as follows: $\ln((1.5)^2)$. Now enter this into your calculator. Press $\boxed{\text{LN}}$ $\boxed{(}$ $\boxed{1}$ $\boxed{.}$ $\boxed{5}$ $\boxed{)}$ $\boxed{x^2}$ $\boxed{)}$ $\boxed{\text{ENTER}}$.

```
ln((1.5)²)
           .8109302162
■
```

There are no special keys to evaluate logarithms in bases other than 10 and $e$ on the calculator. You may however use the fact that $\log_b x = \dfrac{\log x}{\log b} = \dfrac{\ln x}{\ln b}$. Note that you may use either base 10 or base $e$ and you will get the same answer.

For example, to evaluate $\log_4 7$ write it as $\dfrac{\log 7}{\log 4}$ or $\dfrac{\ln 7}{\ln 4}$. Below we see the same solution is given no matter which evaluation is used.

```
log(7)/log(4)
           1.403677461
ln(7)/ln(4)
           1.403677461
```

## Graphing Exponential and Logarithmic Functions

Graphing an exponential or logarithmic function is done in the same manner as the graphs in Chapter 5. However there are some restrictions on the domain and range that assist with choosing the window settings.

**Section 12.1, Example 5** asks for the function $y = \left(\dfrac{1}{2}\right)^x$ to be graphed from -3 to 3. We will graph the function focusing on how to choose window settings instead of using the given setting -3 to 3.

Make sure that if you have a fractional base that you use parenthesis around the fraction. To enter the exponent the caret key will be used. To enter this function into the calculator press [Y=] [(] [1] [÷] [2] [)] [^] [X,T,Θ,n] [ENTER].

Next we will set the window. Start with [ZOOM] [6]. Clearly, negative $y$ values do not occur so Ymin can be set to 0. Because the function decreases so rapidly for negative x, the function does not show up on the screen until $x \approx -4$. Therefore we will set Xmin to -5. If the scale is left at 1 and Ymax is left at 10, we cannot see the function beyond $x = 3$. Therefore, we will set Xmax to 4. The final window settings are X=[-5, 4, 1] and Y=[0, 10, 1]. We then get the graph below.

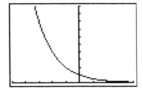

If you want to see that $y = 0$ is a horizontal asymptote, focus in on the positive $x$-axis. Remember that a horizontal asymptote is a line that the function approaches as $x$ gets larger but that the function never touches. Choose the window settings X = [0, 10, 1] and Y = [0, 1, 0.1]. You then get the following graph.

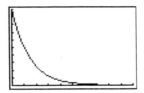

Again, you can see that at $x \approx 7$ you cannot differentiate the $x$-axis from the graph. You could continue in this manner and as you increase Xmax and make the Yscl smaller, you would see that the graph and the $x$-axis do not meet.

In the case where you have exponential growth, to see the asymptote $y = 0$ you would need to focus on the negative $x$-axis. You can also study the rapid growth by increasing Ymax and adjusting your Yscl.

In **Section 12.2, Example 5** asks for the function $y = \log(x)$ to be graphed from 1 to 6. We will graph the function focusing on how to choose window settings instead of using the given setting of 1 to 6.

Enter the function using [Y=] [LOG] [X,T,Θ,n] [)]. Start with the standard window settings by pressing [ZOOM] [6]. Here you can see that $x$ is never negative so change Xmin to 0. Your $x$ values are X=[0, 10, 1]. Since the logarithm grows very slowly, we can reset our $y$ values. Looking at the graph for the current $x$ values, a reasonable choice for the $y$ values is Y=[-2, 2, 1]. You get the following graph.

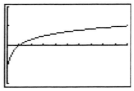

If you want to study the behavior near $x = 0$, try the window settings X=[0, 1, 0.1] and Y=[-2, 0, 0.1]. The graph will show how quickly the function decreases toward negative infinity. Remember that each tick mark on the y-axis represents 0.1 units and only two units in the vertical direction are represented in the graph below.

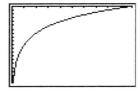

If you want to see how slowly the function grows, use a window such as X=[0, 1000, 100] and Y=[0, 3, 1] as pictured below. Remember that when looking at the x scale, the values go all the way to 1000 and the function's y value just reaches 3.

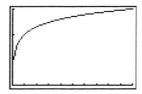

## Checking Solutions to Equations

In past chapters we have seen equations checked numerically and graphically. Exponential and logarithmic equations can be checked using the same techniques.

**Example 4 of Section 12.5** asks that the equation $\log_4 x - \log_4 7 = 2$ be solved. The answer given is $x = 112$. To check this numerically we have $\log_4 112 - \log_4 7 = 2$. The calculator does not work directly in base 4 but these can be re-written as discussed earlier in this chapter: $\dfrac{\log 112}{\log 4} - \dfrac{\log 7}{\log 4} = 2$. Evaluate the left side and we see they are equal.

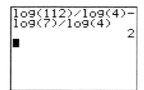

If you want to solve this equation or check the answer graphically, you need to set
$Y_1 = \dfrac{\log x}{\log 4} - \dfrac{\log 7}{\log 4}$ and $Y_2 = 2$ using the $\boxed{Y=}$ button. Use the window settings below. Choosing
the window settings would typically be done by trial and error, slowly increasing the maximum $x$
value, except in this case we know the solution is 112 because we checked it above. Then use
$\boxed{2nd}$ $\boxed{CALC}$ $\boxed{5}$ to locate the intersection which is also your solution.

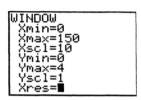 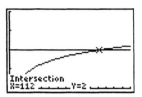

# Chapter 13

# Geometry and Right Triangle Trigonometry

In this chapter of your textbook, all angles are measured in degrees. Therefore you must set your calculator in degree mode. To do this, press $\boxed{\text{MODE}}$. Move the cursor down and highlight DEGREE. Press $\boxed{\text{ENTER}}$. It now should be highlighted as pictured below.

Use $\boxed{\text{2nd}}$ $\boxed{\text{QUIT}}$ to leave this screen.

## Evaluating Trigonometric Functions

Begin by making sure that your graphing calculator is set in degree mode as described above.

**Section 13.6 Exercise 1** asks that $\cos 32.0°$ be evaluated. To evaluate this, press $\boxed{\text{COS}}$ $\boxed{3}$ $\boxed{2}$ $\boxed{.}$ $\boxed{0}$ $\boxed{)}$ $\boxed{\text{ENTER}}$. To evaluate **Exercise 3**, $\tan 24.5°$, press $\boxed{\text{TAN}}$ $\boxed{2}$ $\boxed{4}$ $\boxed{.}$ $\boxed{5}$ $\boxed{)}$ $\boxed{\text{ENTER}}$.

While solving the word problem in **Section 13.7, Example 3**, we must evaluate $x = 4.20 \sin 3.2°$. Here we press $\boxed{4}$ $\boxed{.}$ $\boxed{2}$ $\boxed{0}$ $\boxed{\text{SIN}}$ $\boxed{3}$ $\boxed{.}$ $\boxed{2}$ $\boxed{)}$ $\boxed{\text{ENTER}}$. The results are pictured below.

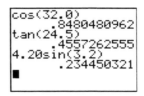

Note that there are no buttons for cotangent, secant and cosecant. To find the values involving these functions the following reciprocal ratios must be used:

$$\csc A = \frac{1}{\sin A}, \quad \sec A = \frac{1}{\cos A}, \quad \cot A = \frac{1}{\tan A}$$

For example, **Section 13.6, Exercise 9** asks for $\cot 76.6°$ to be evaluated. This can be re-written as $\dfrac{1}{\tan 76.6°}$. Enter this into the calculator by pressing $\boxed{1}$ $\boxed{\div}$ $\boxed{\text{TAN}}$ $\boxed{7}$ $\boxed{6}$ $\boxed{.}$ $\boxed{6}$ $\boxed{)}$ $\boxed{\text{ENTER}}$.

An alternative way to look at this is as $(\tan 76.6°)^{-1}$ where the -1 is the exponent and NOT the

inverse discussed below. To enter this you would press $\boxed{\text{TAN}}$ $\boxed{7}$ $\boxed{6}$ $\boxed{.}$ $\boxed{6}$ $\boxed{)}$ $\boxed{x^{-1}}$ $\boxed{\text{ENTER}}$.
Both methods give the same results pictured below.

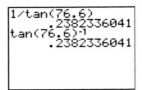

## Inverses of Trigonometric Functions

When you know the value for the trigonometric ratio and are looking for the angle, you use the inverse trigonometry functions on the calculator. These will be accessed using the $\boxed{\text{2nd}}$ key on the calculator followed by the appropriate function. Note that when writing $\sin^{-1}$ this means to take the inverse and is not the exponent -1 which would mean to take the reciprocal.

While solving the word problem in **Section 13.7 Example 2** it is necessary to find $\tan A = \dfrac{5.00}{8.00}$ and $\tan B = \dfrac{8.00}{5.00}$. For the first one press $\boxed{\text{2nd}}$ $[\text{TAN}^{-1}]$ $\boxed{5}$ $\boxed{.}$ $\boxed{0}$ $\boxed{0}$ $\boxed{\div}$ $\boxed{8}$ $\boxed{.}$ $\boxed{0}$ $\boxed{0}$ $\boxed{)}$

$\boxed{\text{ENTER}}$. The answer for angle $A$ is approximately 32 degrees. Remember that your answer is given in degrees because the mode on the calculator is set to degrees. For the second one since it is similar to the first one use $\boxed{\text{2nd}}$ $[\text{ENTRY}]$ to recall the previous calculation and fix the numbers you are dividing. The answer for angle $B$ is approximately 58 degrees.

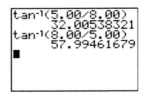

When looking for the inverses of secant, cosecant and cotangent again there are no calculator keys to find these directly. These must first be converted to cosine, sine or tangent respectively. For the conversion use

$$\sin A = \frac{1}{\csc A}, \quad \cos A = \frac{1}{\sec A}, \quad \tan A = \frac{1}{\cot A}.$$

For example, if we work on **Section 13.6, Exercise 19**, we want to find $\alpha$ to the nearest tenth when $\sec \alpha = 1.057$. Using the middle formula above and substituting the value for $\sec \alpha$, we have $\cos \alpha = \dfrac{1}{1.057}$. To find $\alpha$ we now use cosine inverse as follows: $\boxed{\text{2nd}}$ $[\text{COS}^{-1}]$ $\boxed{1}$ $\boxed{\div}$ $\boxed{1}$

$\boxed{.}$ $\boxed{0}$ $\boxed{5}$ $\boxed{7}$ $\boxed{)}$ $\boxed{\text{ENTER}}$. We get approximately 18.9 degrees as shown on the next page.

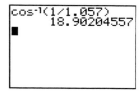

Recall that you may use the round function which is discussed in Chapter 2 of this manual to make sure the decimal is correct.

# Chapter 14

# Oblique Triangles and Vectors

Trigonometric functions with angles that are negative or angles that are larger than 90 degrees are entered into your calculator the same as in Chapter 13. Remember to set your calculator mode to degrees. For directions on setting the mode, see the beginning of Chapter 13 of this manual.

For example if you want to find tan(−30°) press [TAN] [(-)] [3] [0] [)] [ENTER]. The answer is given as -.5773502692.

To find sin(120°) press [SIN] [1] [2] [0] [)] [ENTER]. The answer is given as .8660254038.

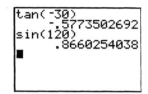

## Inverse Trigonometric Functions Revisited

In Chapter 13, the problems worked with right triangles so the angles encountered were all in the first quadrant and the calculator when using the inverse trigonometric functions always gave us the answer we wanted. In Chapter 14, the problems involve angles that may be more than 90° and if measured counter clockwise may be negative. When working with the inverse trigonometry functions, the calculator will only give you one answer for the angle. You must then use the reference angles as discussed in the main text to find the second angle. Then you use your calculator to make sure the reference angles were used correctly.

The TI-83/83 Plus/84 Plus all use angles between −90° and 90° for sine inverse and tangent inverse. They use angles between 0° and 180° for cosine inverse.

In **Example 14 of Section 14.1**, you are asked to find all angles between 0° and 360° so that the sin θ = 0.342. Make sure your calculator is in degree mode. Press [2nd] [SIN⁻¹] [0] [.] [3] [4] [2] [)] [ENTER]. The angle is given as 19.99877181 which rounds to 20.0 so the first answer is 20°. Since sine is also positive in the second quadrant, we use its reference angle and compute the second angle as 180° − 20.0° = 160.0°. We then check on our calculator that sin(160°) = 0.342 by pressing [SIN] [1] [6] [0] [)] [ENTER]. The calculator shows .3420201433 which when rounded is 0.342. We did not expect to receive an exact answer because we rounded above when finding the first angle.

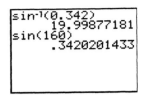

An applied example of where you have to find two angles is given in **Example 4c of Section 14.2**. In this problem, we are asked to find $\sin B = 2/3$. On the calculator we press $\boxed{2nd}$ $\boxed{SIN^{-1}}$ $\boxed{2}$ $\boxed{\div}$ $\boxed{3}$ $\boxed{)}$ $\boxed{ENTER}$. We get approximately 41.8°. Using the reference angle we see that the second angle should be $180° - 41.8° = 138.2°$. We check that sine applied to this angle gives us approximately 2/3. The result below shows that it does.

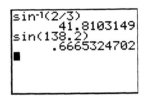

## Law of Sines and Law of Cosines

When using the Law of Sines or Law of Cosines it is very important to get the order of operations and parenthesis correct on the calculator.

In **Section 14.2, Example 3**, we need to calculate $DB = \dfrac{10.2\sin 40.0°}{\sin 5.0°}$. Remember that each time you press $\boxed{SIN}$ that the parentheses open up automatically and you must close it after entering the number. To do this calculation press $\boxed{1}$ $\boxed{0}$ $\boxed{.}$ $\boxed{2}$ $\boxed{\times}$ $\boxed{SIN}$ $\boxed{4}$ $\boxed{0}$ $\boxed{.}$ $\boxed{0}$ $\boxed{)}$ $\boxed{\div}$ $\boxed{SIN}$ $\boxed{5}$ $\boxed{.}$ $\boxed{0}$ $\boxed{)}$ $\boxed{ENTER}$. You will get the answer of approximately 75.2 m.

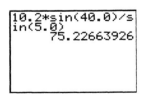

In **Section 14.3, Example 2**, we compute $a = \sqrt{(45.0)^2 + (62.5)^2 - 2(45.0)(62.5)\cos 126.3°}$. Rather than using parenthesis around the numbers we will use the multiplication sign. This will make the parenthesis easier to read. We will write the equation as $a = \sqrt{(45.0^2 + 62.5^2 - 2 \times 45.0 \times 62.5\cos(126.3°))}$. Press $\boxed{2nd}$ $\boxed{\sqrt{\ }}$ to begin the square root. Notice that the parenthesis is automatically opened for you. Then press $\boxed{4}$ $\boxed{5}$ $\boxed{.}$ $\boxed{0}$ $\boxed{x^2}$ $\boxed{+}$ $\boxed{6}$ $\boxed{2}$ $\boxed{.}$ $\boxed{5}$ $\boxed{x^2}$ $\boxed{-}$ $\boxed{2}$ $\boxed{\times}$ $\boxed{4}$ $\boxed{5}$ $\boxed{.}$ $\boxed{0}$ $\boxed{\times}$ $\boxed{6}$ $\boxed{2}$ $\boxed{.}$ $\boxed{5}$ $\boxed{COS}$ $\boxed{1}$ $\boxed{2}$ $\boxed{6}$ $\boxed{.}$ $\boxed{3}$ $\boxed{)}$ $\boxed{)}$ $\boxed{ENTER}$. Notice that at the end, you must press the close parentheses twice. The first one is to close the cosine function and the second one is to close the square root expression. The answer is approximately 96.2.

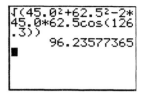

## Vector Addition and Vector Components

To add vectors algebraically they must be resolved into the $x$ and $y$ components, added in the component form to get the resultant and then the magnitude and direction for the resultant must be determined. This is best done by storing the intermediate values on the calculator. Storing the values eliminates the rounding errors introduced when using paper and pencil.

We will do **Section 14.6, Example 3**. We are asked to add vector $A$ which has magnitude 570 and angle 15.0° with the $x$-axis and vector $B$ which has magnitude 350 and angle of 125.0° with the $x$-axis.

Because we cannot use subscripts in the calculator, we will begin by finding the $x$ component of each angle which will be stored in variables **A** and **B** respectively and then add them to find the $x$ component of the resultant vector which will be stored in variable **X**. You will be using the ALPHA key so the following key stroke will be found in green lettering on the keypad.

$A_x$ is found by calculating $A_x = 570\cos(15.0°)$. Press $\boxed{5}$ $\boxed{7}$ $\boxed{0}$ $\boxed{\text{COS}}$ $\boxed{1}$ $\boxed{5}$ $\boxed{.}$ $\boxed{0}$ $\boxed{)}$ $\boxed{\text{STO}\blacktriangleright}$ $\boxed{\text{ALPHA}}$ **[A]** $\boxed{\text{ENTER}}$ to do the calculation and store the result in **A**. Note that the current value of **A** is 550.577721.

Next $B_x$ is found by calculating $B_x = 350\cos(125.0°)$. Press $\boxed{3}$ $\boxed{5}$ $\boxed{0}$ $\boxed{\text{COS}}$ $\boxed{1}$ $\boxed{2}$ $\boxed{5}$ $\boxed{.}$ $\boxed{0}$ $\boxed{)}$ $\boxed{\text{STO}\blacktriangleright}$ $\boxed{\text{ALPHA}}$ **[B]** $\boxed{\text{ENTER}}$. The current value of **B** is -200.7517527.

Then to get the $x$ component of the resultant, these two values must be added. Press $\boxed{\text{ALPHA}}$ **[A]** $\boxed{+}$ $\boxed{\text{ALPHA}}$ **[B]** $\boxed{\text{STO}\blacktriangleright}$ $\boxed{\text{ALPHA}}$ **[X]** $\boxed{\text{ENTER}}$. This sets **X** equal to 349.8259683.

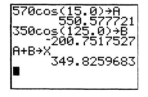

We repeat a similar computation to find the $y$ component and then store it in **Y**. Since we no longer need the $x$ components of the original vectors, we will reuse the memory at **A** and **B**.

To find $A_y = 570\sin(15.0°)$, press 5 7 0 SIN 1 5 . 0 ) STO▶ ALPHA [A]
ENTER. A is 147.5268557.
To find $B_y = 350\sin(125.0°)$, press 3 5 0 SIN 1 2 5 . 0 ) STO▶ ALPHA [B]
ENTER. B is 286.70032155.

Finally add the $y$ components and store them in Y by pressing ALPHA [A] + ALPHA [B]
STO▶ ALPHA [Y] ENTER. Y is 434.2300712.

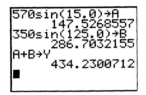

We now have the $x$ and $y$ components of the resultant stored in X and Y respectively. To find the
magnitude of the resultant which is the square root of the resultant's $x$ component squared plus
the resultant's $y$ component squared, press 2nd [√] ALPHA [X] $x^2$ + ALPHA [Y] $x^2$ )
ENTER. The magnitude is 557.6145289.

Finally to find the angle of the resultant vector, you must take the inverse tangent of the
resultant's $y$ component divided by the resultant's $x$ component. Press 2nd [TAN⁻¹] ALPHA [Y]
÷ ALPHA [X] ) ENTER. The angle is 51.14426336 degrees.

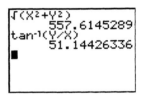

## The Vector Calculator Application

On the TI-83 Plus/TI-84 Plus there is an application that will allow you to complete vector
computations with ease. This application is not available on the TI-83 so you must use the
longer method above.

To access the application press the APPS button located to the right of the MATH button. Scroll
down and highlight SciTools and press ENTER. Press ENTER again to access the application.
Press 4 for VECTOR CALCULATOR.

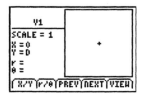

Notice the symbols on the bottom of the screen. To take the corresponding action, press the key on the keyboard directly below your choice.

We will re-do **Example 3 from Section 14.6** using this application.

Press WINDOW for r / θ. Then enter 5 7 0 for r and use the down arrow key ▼ and enter 1 5 for θ. This will enter in the first vector which the calculator will label as Y1.

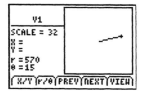

Select **NEXT** by pressing TRACE. Choose r / θ again by pressing WINDOW. Enter 3 5 0 for r and press ▼. Enter 1 2 5 for θ. This vector is entered as Y2.

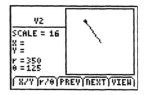

Press TRACE for **NEXT**. Then choose **MATH** (note that this is NOT the math key on the keypad) by pressing GRAPH.

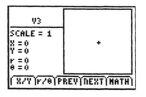

Use **PREV** or **NEXT** until vector Y1 comes up and press GRAPH for **PICK**. Then press Y= for +. Choose **NEXT** by pressing TRACE to get to Y2 and then press GRAPH to **PICK** it.

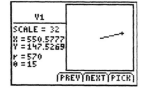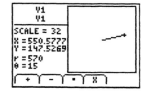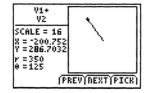

The resultant's $x$ and $y$ components as well as the magnitude and angle are displayed.

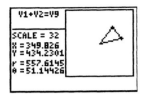

To exit the application use [2nd] [QUIT] repeatedly until you are prompted to EXIT. You then press [Y=] and you will leave the application.

# Chapter 15

# Graphs of Trigonometric Functions

## Angles in Radians and Degrees

The calculator can convert angles from radians to degrees and degrees to radians. See the main text for a discussion of radians.

**Section 15.1, Example 4** asks for 1.75 radians to be converted to degrees. Begin by setting the mode on the calculator to the units you need to end up in, in this case degrees. Press $\boxed{\text{MODE}}$ and arrow down and to the right until the cursor is on the word **DEGREE**. Press $\boxed{\text{ENTER}}$. Use $\boxed{\text{2nd}}$ [QUIT] to exit the mode screen. Next enter the number you wish to convert, $\boxed{1}$ $\boxed{.}$ $\boxed{7}$ $\boxed{5}$. To indicate that this is radians, press $\boxed{\text{2nd}}$ [ANGLE] and choose item $\boxed{3}$ for ʳ or radians. Press $\boxed{\text{ENTER}}$ and you are given that 1.75 radians is approximately 100 degrees.

The second part of this question asks for 270 degrees to be converted to radians. Set the mode to radians, since we want our answer in radians. Press $\boxed{\text{MODE}}$. Use the arrow keys to highlight **RADIAN** and then press $\boxed{\text{ENTER}}$. Remember to use $\boxed{\text{2nd}}$ [QUIT] to exit the mode screen. Then enter the number to convert, $\boxed{2}$ $\boxed{7}$ $\boxed{0}$. Use $\boxed{\text{2nd}}$ [ANGLE] $\boxed{1}$ to indicate degrees. Press $\boxed{\text{ENTER}}$ and you are given that 270 degrees is approximately 4.71 radians.

Note that in the picture below, the answer is given according to the mode in which the calculator was set.

```
1.75ʳ
        100.2676141
270°
          4.71238898
```

## Graphing Trigonometric Functions

Graphing trigonometric functions is very much like graphing any other function. Review Chapter 5 of this manual for general graphing review.

**Example 3 of Section 15.4** works with the equation $y = 4\sin(2x)$. We will be using radians since no unit is indicated. Check the mode setting on your calculator.

To graph $y = 4\sin(2x)$, use [Y=] and enter the function after the Y₁= by pressing [4] [SIN] [2] [X,T,Θ,*n*] [)] [ENTER]. Next the viewing window must be set. There is a special setting for trigonometric functions, found by pressing [ZOOM] [7]. The graph is automatically displayed.

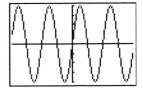

Since the window was set automatically for you, it is necessary to look at the window settings that were chosen to be able to properly read the graph. Simply press [WINDOW] to get them. Notice that in this example the scale for the *x* variable is approximately 1.57 which is $\pi/2$.

```
WINDOW
 Xmin=-6.152285…
 Xmax=6.1522856…
 Xscl=1.5707963…
 Ymin=-4
 Ymax=4
 Yscl=1
 Xres=1
```

Frequently more than one graph is displayed at the same time to see the effects of amplitude, period or displacement. To distinguish between graphs, you may choose to change the line thickness; thin ⋱, normal ⋱, or thick ⋱. Alternatively you can press [TRACE] and use the arrow keys to place the cursor on the graph you are looking at. The equation will be displayed in the upper left hand corner.

In **Section 15.5** the graphs of $y = \sin(x)$ and $y = \sin(x + \pi/2)$ are graphed on the same set of axes. Make sure your mode is set to radians. Enter these into your calculator using [Y=]. Before leaving the screen, move the cursor so that it is over the slash mark in front of Y₂. Press [ENTER] until you see the thick line, ⋱. This means that Y₂ will have the thicker line. Set the window using [ZOOM] [7]. The following graph is displayed.

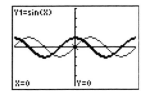

The equation $y = \sin(x + \pi/2)$ is represented by the bolded line. If you are uncertain, press [TRACE] and the equation of the line that the cursor is on, $y = \sin(x)$, is displayed in the upper corner of the screen.

# Chapter 16

# Complex Numbers

## Introduction to Complex Numbers on the Calculator

While the book represents a complex number using the letter $j$, the calculator uses $i$. Locate $i$ on the bottom row of the calculator. You will press [2nd] [$i$] to access it. For this section, we will set our calculator mode to complex numbers by selecting [MODE] and moving the cursor down to REAL, over to a+bi and selecting [ENTER].

Use [2nd] [QUIT] to exit the mode screen.

In **Section 16.1, Example 5** we want to write $-7 + \sqrt{-12}$ as an imaginary number in the form a+bi. After checking that the mode is set to a+bi, enter the equation by pressing [(-)] [7] [+] [2nd] [√] [(-)] [1] [2] [)] [ENTER]. Note that this gives an approximation for the imaginary part of our expression. If you need the exact value, work by hand with the properties of square roots and imaginary numbers to get $-7 + 2i\sqrt{3}$ and check your answer with the calculator. To enter $-7 + 2i\sqrt{3}$, press [(-)] [7] [+] [2] [2nd] [$i$] [2nd] [√] [3] [)] [ENTER]. You see that the approximations are the same.

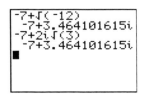

**Section 16.1, Example 6** identifies the real and imaginary parts of a complex number. The calculator does this as well. We will use the complex number menu found by pressing [MATH] and using the [▶] [▶] keys twice so that CPX is highlighted.

To determine the real part of the number $6-i$, press $\boxed{\text{MATH}}$ $\boxed{\blacktriangleright}$ $\boxed{\blacktriangleright}$ $\boxed{2}$ for `real`. We then type the number by pressing $\boxed{6}$ $\boxed{-}$ $\boxed{\text{2nd}}$ $[i]$. Finally press $\boxed{)}$ $\boxed{\text{ENTER}}$. We see the real part is 6. Then to get the imaginary part, press $\boxed{\text{MATH}}$ $\boxed{\blacktriangleright}$ $\boxed{\blacktriangleright}$ $\boxed{3}$ for `imag`. Enter the number; $\boxed{6}$ $\boxed{-}$ $\boxed{\text{2nd}}$ $[i]$. Finally press $\boxed{)}$ $\boxed{\text{ENTER}}$ and we get that the imaginary part is -1.

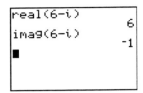

This complex menu also can be used to find the conjugate of a complex number. **Section 16.1, Example 7** asks for the conjugate of $2-7i$. Press $\boxed{\text{MATH}}$ $\boxed{\blacktriangleright}$ $\boxed{\blacktriangleright}$ $\boxed{1}$ for `conj`. Then enter the number by pressing $\boxed{2}$ $\boxed{-}$ $\boxed{7}$ $\boxed{\text{2nd}}$ $[i]$. Press $\boxed{)}$ $\boxed{\text{ENTER}}$ and the conjugate is given as $2+7i$.

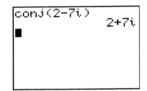

## Arithmetic of Complex Numbers

After setting the mode to `a+bi`, arithmetic on the calculator involving complex numbers is done simply by typing in the desired operation.

In **Section 16.2, Example 2**, $(6+i)+(-2-8i)$ are added. On the calculator press $\boxed{(}$ $\boxed{6}$ $\boxed{+}$ $\boxed{\text{2nd}}$ $[i]$ $\boxed{)}$ $\boxed{+}$ $\boxed{(}$ $\boxed{(-)}$ $\boxed{2}$ $\boxed{-}$ $\boxed{8}$ $\boxed{\text{2nd}}$ $[i]$ $\boxed{)}$ $\boxed{\text{ENTER}}$ and you get $4-7i$. Similarly, in **Example 4**, $(6+i)(-2-8i)$ are multiplied. Press $\boxed{(}$ $\boxed{6}$ $\boxed{+}$ $\boxed{\text{2nd}}$ $[i]$ $\boxed{)}$ $\boxed{\times}$ $\boxed{(}$ $\boxed{(-)}$ $\boxed{2}$ $\boxed{-}$ $\boxed{8}$ $\boxed{\text{2nd}}$ $[i]$ $\boxed{)}$ $\boxed{\text{ENTER}}$ and you get $-4-50i$.

```
(6+i)+( -2-8i )
              4-7i
(6+i)*( -2-8i )
              -4-50i
```

In **Example 9**, we are asked write $\dfrac{1}{2}\left[(1+2i)+\dfrac{1}{1+2i}\right]$ in the standard form $a+bi$. Proper placement of parenthesis is extremely important in a problem like this. Make sure you use parenthesis around the ½ and around the denominator of the fraction 1/(1+2i). Press $\boxed{(}$ $\boxed{1}$ $\boxed{\div}$ $\boxed{2}$ $\boxed{)}$ $\boxed{(}$ $\boxed{(}$ $\boxed{1}$ $\boxed{+}$ $\boxed{2}$ $\boxed{\text{2nd}}$ $[i]$ $\boxed{)}$ $\boxed{+}$ $\boxed{1}$ $\boxed{\div}$ $\boxed{(}$ $\boxed{1}$ $\boxed{+}$ $\boxed{2}$ $\boxed{\text{2nd}}$ $[i]$ $\boxed{)}$ $\boxed{)}$ $\boxed{\text{ENTER}}$.

Then to convert your answer to a fraction, press $\boxed{\text{MATH}}$ $\boxed{1}$ $\boxed{\text{ENTER}}$. You then get your answer $\frac{3}{5}+\frac{4}{5}i$.

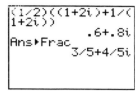

## Rectangular Form and Polar Form Conversions

The TI-83/ 83 Plus/ 84 Plus have the ability to convert the rectangular form of a complex number to polar form and polar form to rectangular form directly. We will use the ANGLE menu shown below. Note that the arrow keys have been use to show the parts of the menu we will be using; functions 5, 6, 7 and 8. You access the menu by pressing $\boxed{\text{2nd}}$ [ANGLE].

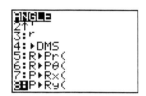

In **Section 16.4, Example 5**, we are asked to express $12.3\angle239.4°$ to rectangular form. The modulus $r$ is 12.3 and the argument $\theta$ is 239.4 degrees.

**Set the mode on your calculator to degrees** since our angle is given in degrees. Then press $\boxed{\text{2nd}}$ [ANGLE] to access the angle menu. Choose $\boxed{7}$ to get the real part of the rectangular form. The function takes two values, the modulus followed by a comma and then the argument. Press $\boxed{1}$ $\boxed{2}$ $\boxed{.}$ $\boxed{3}$ $\boxed{,}$ $\boxed{2}$ $\boxed{3}$ $\boxed{9}$ $\boxed{.}$ $\boxed{4}$ $\boxed{)}$ $\boxed{\text{ENTER}}$. The real part is approximately -6.26. To find the imaginary part, again start by accessing the menu using $\boxed{\text{2nd}}$ [ANGLE] and this time choose $\boxed{8}$. This function takes the same two values so press $\boxed{1}$ $\boxed{2}$ $\boxed{.}$ $\boxed{3}$ $\boxed{,}$ $\boxed{2}$ $\boxed{3}$ $\boxed{9}$ $\boxed{.}$ $\boxed{4}$ $\boxed{)}$ $\boxed{\text{ENTER}}$ again. The imaginary part is approximately -10.6. Putting the two answers together, the rectangular form is $-6.26-10.6i$.

```
P▶Rx(12.3,239.4)
        -6.261209414
P▶Ry(12.3,239.4)
        -10.58712693
■
```

In **Example 6**, we are given the rectangular form $-6.26-10.6i$ and asked to express it in polar form. First we will find the modulus, $r$. Press $\boxed{\text{2nd}}$ [ANGLE] $\boxed{5}$ for the function. We need to enter in the real and imaginary parts of the rectangular form separated by a comma. Press $\boxed{(-)}$ $\boxed{6}$ $\boxed{.}$ $\boxed{2}$ $\boxed{6}$ $\boxed{,}$ $\boxed{(-)}$ $\boxed{1}$ $\boxed{0}$ $\boxed{.}$ $\boxed{6}$ $\boxed{)}$ $\boxed{\text{ENTER}}$. The modulus will be given as approximately 12.3. Next we need to find the argument $\theta$. Press $\boxed{\text{2nd}}$ $\boxed{\text{ANGLE}}$ $\boxed{6}$. Again, we need to enter in the real and imaginary parts of the rectangular form separated by a comma. Press $\boxed{(-)}$ $\boxed{6}$ $\boxed{.}$ $\boxed{2}$ $\boxed{6}$ $\boxed{,}$ $\boxed{(-)}$ $\boxed{1}$ $\boxed{0}$ $\boxed{.}$ $\boxed{6}$ $\boxed{)}$ $\boxed{\text{ENTER}}$. We are given that the argument is approximately -120.6 degrees. This is equivalent to $360 - 120.6 = 239.4$ degrees. Therefore in polar form, we have $12.3(\cos 239.4° + i\sin 239.4°)$.

```
R▶Pr(-6.26,-10.6
)
          12.31046709
R▶Pθ(-6.26,-10.6
)
          -120.5646561
```

# Chapter 17

# Introduction to Data Analysis

Throughout this chapter, the data set of **Section 17.2, Example 2** will be used.  The values are

19, 30, 23, 31, 23, 19, 23, 17, 20, 22, 24, 26, 25, 17, 18, 17,
18, 19, 20, 27, 29, 28, 35, 36, 20, 24, 33, 35, 26, 29, 30, 32

## Creating a Data List

To enter a data list into your calculator, press $\boxed{\text{STAT}}$ to display the STAT menu.

We want to edit our lists so press $\boxed{1}$.  The lists you have will be displayed.  If you want to erase any entries, place the cursor over the entry you want to remove and press $\boxed{\text{DEL}}$.  We will enter the data list above.  Place your cursor in the column labeled $L_1$ and type each number followed by $\boxed{\text{ENTER}}$.

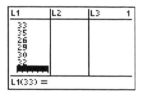

The calculator screen above shows the last 6 data points.  You may scroll up and down through your entries using the arrow keys to see or edit your data.

## Histograms

To create a histogram, the first thing you must do is to enter your data list.  We will use the data list entered above in the section on creating a data list.  The data comes from **Example 2 of Section 17.2**, where we are asked to create a histogram.

Next you must set up for the histogram.  Access the STAT PLOT menu by pressing $\boxed{\text{2nd}}$ $\boxed{\text{Y=}}$.

Select [1] for the first plot. Place the cursor over the word ON and press [ENTER]. Next select the type of graph by placing the cursor over ▫▫ and pressing [ENTER]. Next you must select your list. If L₁ is not showing place the cursor next to Xlist: and press [2nd] [L1] to choose L₁. Finally, make sure Freq: is set to 1. Since when you highlight it you see ▫ flashing, you must first type the [ALPHA] key followed by [1]. Make sure your screen looks like the one pictured below.

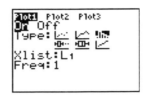

Next press [ZOOM] [9] to display the histogram. If your window is already reasonably set for your data you may just press [GRAPH]. To display the class bounds and the frequency press [TRACE]. On the histogram that corresponds to our data set the first class has a minimum value of 17 and maximum value of 20 1/6. Since our data is all integers, the class contains data values from 17 to 20. There are 11 such data points. To see the data for the other bars, use the arrow keys.

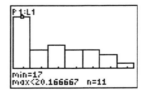

To change the class size press [WINDOW] and change the Xscl to the size you want the class. Xmin is the lowest class bound while Xmax is the largest class bound. For example, since our lowest value is 17, set Xmin = 17 and since our largest value is 36 set Xmax = 36. As displayed above our class size was 3.1666667. If you wanted your class size 2 instead set Xscl =2. When you are done, press [GRAPH] to display the histogram again.

If you have a line through your histogram it is likely that you have a function entered in the [Y=] screen. To disable it go to that screen and place the cursor over any highlighted equals sign and press [ENTER]. To return to the histogram, press [GRAPH].

## Statistics

Section 17.3 demonstrates how to calculate the mean and the median. Section 17.4 works with the standard deviation and range. The calculator can do all of these using the LIST MATH menu accessed by pressing [2nd] [LIST] [▶] [▶].

To do any of these calculations your data list must be entered into the calculator. We will use the data set (Example 2 of Section 17.2) entered in the section on creating data lists.

To find the mean of our data we press [2nd] [LIST] [▶] [▶] [3] [2nd] [L1] [)] [ENTER].

To find the median of our data set, press [2nd] [LIST] [▶] [▶] [4] [2nd] [L1] [)] [ENTER].

To find the standard deviation of the data set, press [2nd] [LIST] [▶] [▶] [7] [2nd] [L1] [)] [ENTER].

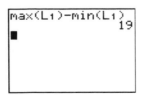

To find the range, we must find the maximum value and subtract the minimum value. Press [2nd] [LIST] [▶] [▶] [2] [2nd] [L1] [)] [−] [2nd] [LIST] [▶] [▶] [1] [2nd] [L1] [)] [ENTER].

```
max(L1)-min(L1)
               19
■
```